Science and Selection

One way to understand science is as a selection process. David Hull, one of the dominant figures in contemporary philosophy of science, sets out in this volume a general analysis of this selection process that applies equally to biological evolution, the reaction of the immune system to antigens, operant learning, and social and conceptual change in science. Hull aims to distinguish those characteristics that are contingent features of selection from those that are essential.

Part One focuses on selection in biological evolution. Part Two contains essays that treat science itself as a selection process. Hull explores questions such as: How are scientists able to cooperate so extensively in such competitive situations, why are scientists so much better than members of other professions at policing themselves, and how come science is so clearly progressive? The answers given to these questions are intended to be themselves scientific. Hence anything that Hull's theory implies about scientific theories as such should apply to his theory as well. In Part Three, Hull examines the testing of our views about science, arguing that if testing plays such a crucial role in other areas of science, then it must play the same role in the study of science.

Science and Selection brings together David Hull's important essays on selection (some never before published) in one accessible volume. It will be of interest to students and professionals in philosophy of science and evolutionary biology, and any others interested in the study of science.

David L. Hull is Dressler Professor in the Humanities at Northwestern University.

CAMBRIDGE STUDIES IN PHILOSOPHY AND BIOLOGY

General Editor
Michael Ruse   *University of Guelph*

Advisory Board
Michael Donoghue   *Harvard University*
Jean Gayon   *University of Paris*
Jonathan Hodge   *University of Leeds*
Jane Maienschein   *Arizona State University*
Jesús Mosterín   *Instituto de Filosofía (Spanish Research Council)*
Elliott Sober   *University of Wisconsin*

Other Titles
Alfred I. Tauber: *The Immune Self: Theory or Metaphor?*
Elliott Sober: *From a Biological Point of View*
Robert Brandon: *Concepts and Methods in Evolutionary Biology*
Peter Godfrey-Smith: *Complexity and the Function of Mind in Nature*
William A. Rottschaefer: *The Biology and Psychology of Moral Agency*
Sahotra Sarkar: *Genetics and Reductionism*
Jean Gayon: *Darwinism's Struggle for Survival*
Jane Maienschein and Michael Ruse (eds.): *Biology and the Foundation of Ethics*
Jack Wilson: *Biological Individuality*
Richard Creath and Jane Maienschein (eds.): *Biology and Epistemology*
Alexander Rosenberg: *Darwinism in Philosophy, Social Science and Policy*
Peter Beurton, Raphael Falk and Hans-Jörg Rheinberger (eds.): *The Concept of the Gene in Development and Evolution*
James G. Lennox: *Aristotle's Philosophy of Biology*

Science and Selection

Essays on Biological Evolution and the Philosophy of Science

DAVID L. HULL
Northwestern University

CAMBRIDGE
UNIVERSITY PRESS

PUBLISHED BY THE PRESS SYNDICATE OF THE UNIVERSITY OF CAMBRIDGE
The Pitt Building, Trumpington Street, Cambridge, United Kingdom

CAMBRIDGE UNIVERSITY PRESS
The Edinburgh Building, Cambridge CB2 2RU, UK
40 West 20th Street, New York, NY 10011-4211, USA
10 Stamford Road, Oakleigh, VIC 3166, Australia
Ruiz de Alarcón 13, 28014 Madrid, Spain
Dock House, The Waterfront, Cape Town 8001, South Africa

http://www.cambridge.org

First published 2001

Printed in the United States of America

Typeface Times Roman 10.25/13 pt.   *System* QuarkXPress [BTS]

A catalog record for this book is available from the British Library.

Library of Congress Cataloging in Publication Data
Hull, David L.
Science and selection : essays on biological evolution and the philosophy of science /
David L. Hull.
p.   cm. – (Cambridge studies in philosophy and biology)
Includes bibliographical references (p.).
ISBN 0-521-64339-2 (hb) – ISBN 0-521-64405-4 (pb)
1. Evolution (Biology) – Philosophy.   2. Science – Philosophy.   3. Natural
selection – Philosophy.   I. Title.   II. Series.

QH360.5 .H86   2000
576.8'01 – dc21
                                                                    00–027938

ISBN 0 521 64339 2 hardback
ISBN 0 521 64405 4 paperback

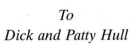

To
Dick and Patty Hull

Contents

Contents

Introduction

Referring to the "evolution of science" is easy enough just so long as all one means by this appellation is that science changes. Lots of things change. The crucial issue is the mechanisms that produce this change. What sorts of mechanisms play what sorts of roles in conceptual change in science? Are they in a significant sense similar to the sorts of mechanisms that function in gene-based biological evolution? An increasingly common way to view scientific change is as one more instance of a selection process. Just as biological species evolve primarily through variation and differential perpetuation, conceptual systems in science change in the same way. In addition, the reaction of the immune system to antigens and operant learning are also referred to frequently as involving selection processes. Are all four of these systems instances of a more general sort of process or is each *sui generis*? A high proportion of students interested in evolutionary epistemology reason from selection in gene-based biological evolution to selection in general, resulting in certain peculiarities of biological selection being elevated to general features of selection. In this collection of papers, I compare all four sorts of selection to see if they have enough in common to count as the same sort of process. Can a general analysis of selection be set out which applies equally to all four sorts of selection?

No one seems to object to the functioning of the immune system as being a selection process. More doubts are likely to be raised to operant learning in part because of doubts about the adequacy of this psychological theory, but in the past few years operant psychologists have made considerable headway in their research program, philosophical objections to it to one side. Even stronger objections have been raised to viewing science as a selection process. Most of these objections stem from a simplistic understanding of gene-based biological evolution. So

1

the story goes: in biological evolution, the basic units of replication are discrete genes of the same size, kind, and importance, while in conceptual evolution, no such discrete units can be found. But anyone who knows anything about genetics knows that the preceding characterization is a serious oversimplification. If caricatures of selection in gene-based biological evolution are taken as standard, then none of the other putative instances of selection count as selection either – not even the functioning of the immune system.

In the past the most obvious way to organize living entities was into a hierarchy of genes, organisms, and species. My best guess is that if we are to have a version of evolutionary theory that is adequate for biological evolution, we will need much more general terms than these. My suggestion is that this traditional way of viewing the living world be replaced by *replicators*, *interactors*, and *lineages*. All three notions are much more general than the commonsense notions that they are designed to replace. They are general enough to accommodate gene-based biological evolution. They are also general enough to include the reaction of the immune system to antigens and operant learning. With only minor modification they are also adequate for conceptual evolution.

I would not expect evolutionary biologists to adopt this alternative way of conceptualizing the living world if all it did was allow "evolution via selection" to be applicable to conceptual change, but I think that it solves or at least dissolves some persistent problems in evolutionary theory as a strictly biological theory. For example, biologists have long argued over the level at which selection occurs. On my account, selection is not one process but two intricately connected processes – replication and environmental interaction. Replication occurs primarily at the level of the genetic material. Interaction occurs at numerous levels from genes, cells, and organisms to colonies, demes, and possibly entire species. Replication is necessary for selection; it is not sufficient. Environmental interaction is also necessary for selection; it too is not sufficient. Any account of selection must make reference to *both* of these processes. There are units of replication and units of environmental interaction. They are the primary units. Units of selection also exist but only in a derivative and highly misleading sense. For good *biological* reasons, the hierarchy of genes, organisms, and species needs to be made more general.

If selection is characterized in terms of genes, organisms, and species, it certainly does not apply to conceptual change. Although a few basic

2

behavioral dispositions may be passed along in the genes, for most of conceptual evolution changes in gene frequencies play no role whatsoever. During the past couple of centuries, a tiny percentage of people have studied the evolutionary process and have come to understand it reasonably well. The range of gene frequencies that they exhibit does not differ appreciably from the rest of humanity. Our understanding of the role of genes in selection is not in any sense "in our genes." If one takes the traditional gene/organism/species perspective, conceptual evolution seems "Lamarckian" while gene-based biological evolution does not. In addition, if the traditional perspective is adopted, conceptual evolution occurs much faster than biological evolution. People can change their conceptual systems several times over in a single generation. So the story goes: cross-lineage borrowing is impossible in biological evolution but is common in conceptual change. Biological evolution is characterized by the dominance/recessiveness relation, while nothing like it occurs in conceptual evolution. Biological evolution is Mendelian, while conceptual change is not.

The preceding differences between biological and conceptual change turn out to be illusory. At most they are differences in degree, not kind; for example, hybridization in organisms turns out to be very common if plants are considered organisms. If the generation times of such long-lived organisms as vertebrates are taken as basic, then evolution occurs very slowly relative to conceptual change. However, most organisms reproduce much more quickly than these "paradigm" organisms. Some organisms (such as viruses and bacteria) reproduce at rates that put conceptual evolution to shame. But more importantly, physical time is the wrong measure for selection processes. For replication the number of generations is what counts, even if elephants take much longer to reproduce than do slime molds and blue-green algae. Given physical time, some lineages change very slowly; others with dizzying speed. Certainly, a belief in the evolution of species has had a long, slow history, while cell phones are quickly taking over the world.

In general, many of the disanalogies raised by critics of treating conceptual change as a selection process stem from a misunderstanding of biological evolution. Others arise from the inability to entertain alternative perspectives. Old-think retains a powerful grip on all of us. The issue is not whether the new perspective that I am urging is *better* than the traditional perspective but that many people seem incapable of even *understanding* this new alternative. It just don't sound right (see

3

Piattelli-Palmarini 1986). Of all the objections that have been raised to viewing conceptual change in terms of selection and evolution, concerns about intentionality and agency seem the most legitimate. To be sure, intentionality plays a much more pervasive role in conceptual change than it does in the immune system or in gene-based biological evolution, but agency introduces issues more fundamental than just intentionality (for efforts to program computers by introducing random changes and selecting the results, see Taubes 1997).

In early versions of the papers in this collection, I claimed that scientists are important interactors in scientific change. As several commentators have pointed out to me, this is a mistake. Scientists certainly enter into conceptual change in science, but as agents that facilitate interaction, not as interactors themselves. Just about the only changes that I have made in the papers included in this collection are the elimination of references to scientists as interactors and the replacement of "produce" in my original definition of "selection" (Hull 1980) with the more general "cause" (Hull 1988a, 1988b).

In Part I of this book, I concentrate primarily on the distinction between replication and environmental interaction as it functions in biological evolution. Are there good biological reasons for adopting this way of thinking about selection with respect to biological evolution? I also ask this same question with respect to two additional examples of selection that do not pose the same apparent array of problems as does conceptual change – the reaction of the immune system to antigens and operant learning. However, any serious attempt to compare these instances of selection requires a deep understanding of the processes involved, so deep that I could not kid myself that I possessed it for immunology and operant behavior. For this reason, I joined forces with an immunologist, Rod Langman, and a behavioral psychologist, Sigrid Glenn. I am grateful to them for allowing me to include our joint paper in this anthology.

In Part II I turn to a more extensive investigation of conceptual change in science as a selection process. However, I do not think that conceptual change in science can be totally explained in terms of a single, universally distributed mechanism unique to conceptual change in science. Additional mechanisms play significant roles in biological evolution. Why not conceptual evolution? Nor are the selective mechanisms that function in conceptual evolution universally distributed in science. They are more prevalent in some areas of science than in others. These mechanisms are also not unique to science. They can be found in certain areas outside of science (depending, of course, on how broadly one defines "science"). But I am committed to the view that science works best at realizing its traditional goals when the selection processes that I set out are most prevalent, and that these mechanisms play a much more important role within science than outside science. In particular, scientific claims can be "tested" much more directly and easily than in other areas of human endeavor. This is not to say that scientific hypotheses can be directly and easily tested. They cannot be. Testing is extremely difficult, no matter the area.

In the papers on selection in science included in this collection, I make two pairs of fundamental distinctions. One is between groups of scientists and their conceptual systems; for example, between the Darwinians and Darwinism. Science is clearly a social process. It occurs in various societies and exhibits its own social organization. In this book I all but ignore the issue of larger social connections. Do male scientists produce male science just because they are male? I have my doubts, but I am much more confident about the effects on science of the social organization of science. Primatologists tell us that hamadryas baboons are organized in increasingly large groups to perform successively more general functions. Scientists exhibit a comparable organization. The smallest groups are research teams made up of a dozen or so researchers, pooling their conceptual resources as they work in close proximity to each other – the way that Glenn, Langman, and I have done in the paper included as Chapter 3.

More inclusive groups have been aptly termed "invisible colleges." They consist of all the scientists contributing to a particular literature. More inclusive groups of scientists, such as all solid-state physicists, tend to be increasingly amorphous and less operative in the ongoing process of science. The important feature of these groups of scientists is that they be defined *socially* in terms of such relations as coauthoring papers and exchanging e-mail messages, not conceptually in terms of sharing common beliefs. They *may* share such beliefs, but they need not. If these groups are followed through time, their membership waxes and wanes. In most instances, they last for only a dozen or so years and then merge back into the larger scientific community. A few succeed in taking over their respective fields. In either case, they cease to exist as small, semi-isolated research groups.

Science is also a conceptual process. If the course of science is to be understood, conceptual systems such as Mendelian genetics and ther-modynamics have to be identified. Just as research groups as social groups have to be distinguished in terms of social relations, conceptual systems have to be individuated conceptually. Certain beliefs tend to go together. If one samples a particular area of science at any one time, beliefs are not distributed continuously. They tend to clump. Certain evolutionary biologists think that species evolve gradually under the twin influence of selection and variation and these variations are unre-lated to the needs of the organisms that exhibit them. Others accept variation and selection but think that evolution occurs in spurts. If such conceptual systems are followed through time, they too undergo change. If "Darwinism" is followed through time, certain tenets become more prevalent while others disappear. In short, both research groups and conceptual systems form historical entities. One reason to dis-tinguish between social groups and conceptual systems in science is that they interact in extremely complex ways. If those of us who study science fail to make this distinction, everything seems to run into everything.

Thus, the first conceptual distinction that I utilize in the following papers is between socially defined scientific groups (e.g., the Darwini-ans) and conceptually defined conceptual systems (e.g., Darwinism). The second is between conceptual inclusive fitness and the demic struc-ture of science. Both notions are borrowed from evolutionary biology. Organisms can pass on copies of their genes directly via having descen-dants, whether or not this reproduction is sexual or asexual. In some organisms, however, kin selection occurs. An organism can pass on copies of its genes directly, but it can also increase the frequency of copies of these same genes indirectly through helping its close relatives. An organism's inclusive fitness is the summation of all the effects that it has on subsequent generations.

Scientists behave in much the same way. They can introduce their conceptual innovations into science and pass them on directly to later generations, but they can also increase the frequency of their contri-butions by aiding one another. In the following papers I cite my own work quite frequently, but I also cite the work of others who cite my work, and so on. Citing is important, but the most fundamental causal factor in science is *use*. Scientists use each other's work and thereby imply its value. One nice thing about science is that only scientists functioning as scientists can use each other's work. In the following papers I set out in some detail the implications of this fundamental character of the social structure of science and its effects on conceptual change. One of the most basic processes that characterize science is appropriately termed "invisible hand." Scientists tend to behave in ways that are calculated to increase their own conceptual inclusive fitness. The credit system in science is so structured that in doing so they also fulfill the explicit goals of science. What is good for General Motors is not always good for the country, but more often than not, what is good for individual scientists is good for science. In science invisible hands join with invisible colleges to make science work the way that it does.

Finally, one problem with biological evolution is that, as slow as it proceeds from the human perspective, it occurs much more quickly than most mechanisms can explain. One solution has been to pay attention to the effects of small groups. Given huge numbers of organisms, the amount of time it takes to change gene frequencies is prohibitive. However, if a mutation occurs in a small population that is effectively isolated from the rest of its species, the frequency of a new gene can be increased quite rapidly, perhaps resulting in fixation within this small group. Then that group can expand relative to other groups. For example, blue eyes in humans have been around since recorded time, probably much, much earlier. Yet currently less than 1 percent of the human population exhibits blue eyes. Blue eyes are relatively prevalent in Scandinavia while they are all but absent from the huge populations living throughout the rest of the world. As I discussed above, a significant proportion of scientists belong to small scientific research groups. New ideas have a better chance of becoming accepted in such small groups than in more inclusive scientific groups. Scientists working in such conceptual demes should be much more productive than scientists working in relative isolation – and they are.

Throughout most of its history, philosophy has been defined in such a way that evidence cannot possibly bear on it. Increasingly, however, those of us who are philosophers by training have begun to interpret philosophy in such a way that evidence does bear on the views that we express. Testing strictly scientific hypotheses is very difficult. Testing claims about science is even more difficult, but test them we must. Given the set of views expressed in this volume, all sorts of things about science follow. As mentioned above, scientists who are parts of relatively small research groups should be more innovative and effective than scientists working in isolation. Scientists should behave in ways to increase their inclusive fitness. Scientists certainly want credit for their contributions, but they also need support. Coming up with new ideas is very difficult. Getting other scientists to incorporate these views into their own work is even more difficult. What strategies do scientists use to accomplish these ends? Are they compatible with the general theory of science that I have set out? In Part III of this anthology, I discuss how views about science can be tested empirically.

A recurrent demand made of anyone attempting to develop a theory about how conceptual systems evolve is that we set out in advance the basic units of this process, but this demand is too strong. Certainly authors of no other theory in the history of science had to set out in advance the basic units that function in the processes being postulated. The scope and nature of these units emerge only as the theory is articulated and tested. For early investigations, something as simple and discrete as the appearance of cladograms in the literature can serve as a rough first approximation to the units of replication. These schematic branching diagrams can be spotted without reading hundreds of papers in dozens of journals. If one sticks to those journals that deal with biological classification and the inferring of phylogenies, very few borderline cases occur. Either a paper includes cladograms or it does not. Hence the spread of cladograms can be charted in the scientific literature.

Numerous arguments have been published to show that viewing conceptual evolution in terms of selection is in principle impossible. I put very little stock in such in-principle arguments. Time and again workers who are apparently incapable of understanding how conclusive these a priori arguments are succeed in doing the impossible. For example, see Griesemer and Wimsatt (1989) for a study of the evolution of Weismann diagrams in the scientific literature and Pocklington and Best (1997) for a method of individuating conceptual units that can function in conceptual selection. Stone (1996) provides a cladistic analysis of mathematical models used to analyze and classify various sorts of shells. One thing that he discovered about these mathematical models is that they include numerous instances of cross-lineage borrowing, but he argues that this feature of conceptual change does not differ appreciably from ordinary biological evolution. Lots of cross-lineage borrowing occurs in biological evolution as well.

In sum, this collection of papers is divided into three parts. In Part I I set out a general analysis of selection with respect to biological evolution and see how well it applies to the immune system and operant learning. In Part II I apply this same analysis to conceptual change in science. Finally, in Part III I discuss how the claims just made can be tested. I do not include much of the relevant data, which has been published elsewhere (see summaries in Hull 1988a).

I

Selection in Biological Evolution

1

Interactors versus Vehicles

The distinction between organisms and species is as old as Western thought. Organisms are discrete, well-organized bodies that go through life cycles and die, while species are groups of similar organisms that mate and produce equally similar offspring. In 1859 Darwin added an evolutionary dimension to both concepts. According to Darwin, organisms are the things that possess the adaptations that allow some of them to cope better with their environments than do other organisms. Some organisms live long enough to reproduce; others do not. Through the culling action of selection, later generations can depart significantly in their characteristics from earlier generations. As a result, species evolve. In this century, genes joined organisms and species to form the basis for our common conceptions of biological phenomena. Genes are discrete bodies arranged linearly on chromosomes. They code for the structure of organisms and are passed on in reproduction. All that is needed to fit genes into an evolutionary framework is to note that on occasion they mutate.

As neat and intuitively appealing as the preceding characterization may be, biologists are challenging every part of it. Some biologists insist that the only entities that need to be referred to explicitly in evolutionary theory are genes. At bottom, evolution is a function of alternative alleles gradually replacing one another. Evolution is nothing but changes in gene frequencies. Other biologists insist that organisms are the primary focus of selection, and that individual genes cannot be selected in isolation from the effects of the entire genome in the production of organisms. Still others maintain that entities more inclusive than single organisms can be selected – possibly even species themselves. Others insist that selection is not as important to evolutionary change as advocates of the synthetic theory think, and that

13

other factors are responsible for many if not most of the changes that occur.

Many of the issues that divide present-day evolutionary biologists are largely empirical, e.g., the prevalence of more gradualistic versus more saltative forms of evolution, the amount of genetic material that plays no role in the production of phenomes, and the extent of genetic disequilibrium. Others stem from the way in which biological phenomena are conceptualized. In this chapter I concentrate on conceptual issues, in particular the way that the traditional gene/organism/species hierarchy has influenced how evolutionary biologists conceive of the evolutionary process. Throughout the history of science, the ways in which scientists have conceived of natural phenomena have influenced the results of empirical research in ways that could not have been anticipated. The story of the tortoise and the hare is only one example. On some very commonsense notions of space and time, the hare should never be able to catch the tortoise; however, it does. Organisms and species are no less commonsense conceptions, conceptions that continue to bias how we all view biological evolution. These biases, in turn, bias how we view conceptual evolution when it is interpreted as an evolutionary process. Behavior evolved as surely as any other phenotypic characteristics of organisms and should be explicable in the same general terms – if they are general enough. Organisms can learn about their environments from interacting with them. What is more interesting, they can pass on this knowledge. They can learn from one another. Social learning has been developed to its greatest degree in science. Might not social learning in general and science in particular be explicable in these same terms? Might not biological evolution and conceptual change both be selection processes? If so, then we are aware of three different sorts of selection processes: biological evolution, the reaction of the immune system to antigens, and learning.

Although the source of a view is irrelevant to its ultimate validity, certain perspectives in the history of science have such bad track records that the presence of one of them in a conceptual system should at least raise doubts about the system. Anthropocentrism has long been recognized as an evil in science, a bias that supposedly was shed centuries ago. Yet it continues to influence the way we conceptualize the evolutionary process.

As organisms go, human beings are quite large, well organized, and discrete in space and time. We also reproduce sexually and give rise to our young in such a way that parents and offspring are easily distinguishable. Our reproduction and our growth are quite different processes. Similar observations hold for nearly all the organisms that immediately spring to mind when we think of organisms. The paradigm of an organism is an adult vertebrate, preferably a mammal. Unfortunately, these paradigmatic organisms are at the tail end of several important distributions. The vast majority of organisms that have ever lived have been small, unicellular, and asexual. According to recent estimates, systematists have described nearly 1.7 million species of organisms. Of these, about 751,000 are insects, 250,000 are flowering plants, and only 47,000 are vertebrates. But nearly all vertebrate species have been described, while most species of insects remain undescribed. According to one estimate, 30 million insect species are probably extant. But even that number shows a bias, because it includes only extant organisms when easily 99 percent of the species that have ever lived are extinct. Roughly 3.5 billion years ago, life evolved here on Earth. Not until 1.3 billion years ago did eukaryotes evolve. None of these were large, multicellular organisms, nor did they reproduce sexually. Multicellularity and sexuality evolved only 650 million years ago, during the Precambrian era. Hence, it seems strange to pick even insects as the paradigmatic organism, let alone vertebrates. The most common organisms that ever existed are blue-green algae.

None of this would matter to science if similar biases did not influence how evolutionary biologists think of biological evolution. When we think of evolution, we tend to think of fruit flies, flour beetles, deer, and humans. We do not think of slime molds, corals, dandelions, and blue-green algae, but if evolutionary theory is to be truly adequate it must apply to all sorts of organisms, not just to those organisms most like us. Multicellularity and sexuality are rare, peculiar, aberrant, deviant, yet nearly all the literature of evolutionary biology concerns large, multicellular organisms that reproduce sexually, and almost none of it deals with the vast majority of organisms. Critics complain of those biologists who want to generalize from the evolution of ordinary phenotypic traits to the evolution of behavior, but we have yet to generalize our understanding of the evolutionary process to the ordinary phenotypic traits of most of the organisms that have lived. If we are not sure whether our current understanding of biological evolution applies unproblematically to reproduction in blue-green algae, perhaps we should be a bit cautious about generalizing to the social organization of African hunting dogs or Yānomamö Indians. To put this cautionary note differently: one should not dismiss cultural or conceptual evolution as aberrant on the basis of such peculiar phenomena as the transmission of eye color in fruit flies. Perhaps a theory of evolution that would be adequate to handle the entire range of organisms that have functioned in this process might also be adequate to handle cultural and conceptual evolution.

Only a very few biologists have protested the biases so inherent in the literature of evolutionary biology (e.g., Bonner 1974; Thomas 1974; Janzen 1977; Dawkins 1982a,b; Jackson et al. 1986). They complain that many organisms lack *all* the characteristics usually attributed to organisms. Some organisms are not very well organized, at least not throughout their entire life cycle. For example, organisms that undergo considerable metamorphosis become dedifferentiated between stages, losing all their internal organization. In such circumstances, the parts of an organism can be rearranged quite extensively without doing much damage. Nor are the spatiotemporal boundaries of all organisms especially sharp. Some organisms go through stages during which they dissolve into separate cells. It becomes all but impossible in such circumstances to decide where one organism begins and another ends, whether one organism is present or hundreds. Zoocentrism notwithstanding, plants are organisms too. Furthermore, a strawberry patch may look like a series of separate plants until we notice the runners that connect those plants into a single network.

As foreign as these conceptions are to zoologists, botanists recognize tillers and tussocks, ramets and genets. For example, grasses frequently grow in tufts (tussocks) composed of numerous sprouts (tillers) all growing from the same root system. Which is the plant: each tiller, or the entire tussock? More generally, botanists term each physiological unit a ramet, and all the ramets resulting from a single zygote a genet. According to Harper (1977), natural selection acts on the genet, not on the ephemeral ramets. As Cook (1980, pp. 90–1) remarks: "Through the eyes of a higher vertebrate unaccustomed to asexual reproduction, the plant of significance is the single stem that lives and dies, the discrete, physiologically integrated organism that we harvest for food and fiber. From an evolutionary perspective, however, the entire clone is a single individual that, like you or me, had a unique time of conception and will have a final day of death when its last remaining stem succumbs to age or accident."

None of this would matter much if the organismic level of organization did not exercise such a disproportionately strong influence over the way in which evolutionary biologists tend to conceptualize the evolutionary process. If selection is a process of differential perpetuation of the units of selection, and if organisms are the primary focus of selection, then we had better know which entities we are to count – e.g., whether to count each little tuft of crabgrass in a field or the entire field. In cases of sexual reproduction, the distinction between reproduction and growth is usually quite clear and can be used to distinguish organisms. Offspring tend to be genetically quite different from their parents and siblings, and the genetic differences can aid in distinguishing separate organisms. But in cases of asexual "reproduction," our commonsense conceptions begin to break down once again. If the two cells that result from mitosis stay in physical contact with each other, we tend to think of them as parts of a single organism and to count the instance of mitotic division as growth. However, if the daughter cells float away from each other we treat them as separate organisms and count the instance of mitotic division as reproduction. Thus, the distinction between growth and reproduction that makes so much sense for "higher" organisms makes little sense in such cases. Why is continued physical contact so important? As long as runners continue to connect all the various strawberry plants in a patch into a single network, is it to count as a single organism? If one of these runners is severed, are there suddenly two organisms? As always, common sense is not good enough for the needs of science. (For one set of answers to the preceding questions, see Dawkins 1982a.)

Precisely the same sorts of problems arise at the genetic level. Early geneticists extrapolated from conceptions of macroscopic entities to genes. Genes, they thought, were like beads on a string. As genetics continued to develop and was eventually joined by molecular biology, we discovered that genes are not in the least like beads on a string. Only in very special circumstances can we treat single genes as if they controlled discrete characters. Epistatic effects are too common. Nor is the genome a crystalline lattice. Instead it seethes with activity: genes turning on and off, introns being snipped out, other segments moving from place to place in the genome, and so on. Even though all this turmoil at the genic level may have very little to do with adaptive phenotypic change (King 1984), it cannot be ignored in the individuation of genes. Although evolutionary biologists disagree about the sufficiency of genes for an adequate characterization of the evolutionary process, they all agree about their necessity. If changes in gene frequencies play an essential role in evolution, then we had better be able to count genes.

Williams's (1966, p. 25) solution to the aforementioned complexities is to individuate genes in terms of selection. An evolutionary gene is "any hereditary information for which there is a favorable or unfavorable selection bias equal to several or many times its rate of endogenous change." Just as the limits of organisms are highly variable once one acknowledges the existence of such "nonstandard" organisms as dandelions and slime molds, the limits of evolutionary genes are also highly variable, depending on several contingent factors such as frequency of crossover. In organisms that reproduce sexually, the evolutionary gene tends to be quite small. In cases of asexual reproduction, it can be the entire genome. On the definition urged by Williams (1966) and adopted by Dawkins (1976), the genomes of some organisms consist in hundreds of thousands of genes; those of others in only one. Thus, from the perspective of either Mendelian or molecular genes, evolutionary genes are highly variable in size. Conversely, from the evolutionary perspective, Mendelian and molecular genes are no less artificial chimeras.

The most frequently voiced objection to Williams's evolutionary definition of *gene* is that it precludes neutral genes by definitional fiat. But this objection is no objection at all, because comparable implications follow from any definition in terms of activity. For example, Mendelian genes are defined in terms of patterns of phenotypic transmission. The only genes that count as Mendelian genes are those that exhibit phenotypic variation. If there are no alleles, there are no Mendelian genes. Of course, there are segments of the genetic material that do not have any differential effect on the phenotype. They are no less a part of the genetic material even though they do not function as Mendelian genes. Only if one thinks that a particular gene concept must subdivide all the genetic material into units of some sort or other do the preceding observations count as objections. If Williams's evolutionary gene concept must be rejected because it distinguishes only those genes that enter differentially into the evolutionary process, then the concept of the Mendelian gene must be rejected as well because it distinguishes only those genes that enter differentially into intergenerational character transmission.

Parallel problems arise at the third level of our commonsense biological conceptions, the species level. Given our relative size, duration, and perceptual acuity, organisms seem to be highly structured, discrete individuals; species do not. On the contrary, species appear to be little more than aggregates of organisms. Through the years biologists have chipped away at this bit of common sense as well (Dobzhansky [1937] 1951, pp. 576–80; Mayr 1963, p. 21; Hennig 1966, p. 6). However, it was Ghiselin (1966, 1974, 1981) who finally forced biologists to recognize that species as units of evolution are not "mere classes" or "just sets" but are more like individuals. Species certainly do not exhibit anything like the structure presented by the most highly organized organisms; however, they do possess spatiotemporal characteristics, and some even exhibit what is commonly termed "population structure." According to Michod (1982, p. 25), "population structure" refers traditionally to "any deviation from panmixia resulting in nonrandom association between genotypes during mating." But Michod sees no reason not to extend this term to include nonrandom associations during any part of a life cycle. I agree with Kitcher (1984) and Williams (1985) that removing or rearranging parts is likely to have a more serious effect on most organisms than on kinship groups, populations, or species, but the differences are in degree rather than in kind. The distinction that is commonly drawn between well-organized, discrete organisms and these more inclusive entities is not as absolute as one might think.

From these and other considerations, numerous authors have argued that species are the same sort of thing as genes and organisms – spatiotemporally localized individuals. Certainly species do not *seem* to be the same sort of thing as genes and organisms when one thinks of genes as beads on a string and vertebrates as typical organisms; however, once one surveys the wide variety of entities that count as genes and organisms, the suggestion begins to look more plausible. More important, this shift in our conception of species matters. It influences in fundamental ways the manner in which we understand the evolutionary process (Hull 1976, 1978a; Eldredge and Salthe 1984; Vrba and Eldredge 1984; Eldredge 1985). For instance, if species are conceived of as the same sort of things as genes and organisms, it is at least *possible* for them to perform the same functions in the evolutionary process. For instance, if species are conceptualized as individuals, it is at least possible for them to be selected. It does not follow, of course, that they are (Sober 1984). Ghiselin (1985, p. 141) presents this point as follows: "It would seem that species do very few things, and most of these are not particularly relevant to ecology. The speciate, they evolve, they provide their component organisms with genetical resources, and they become extinct. They compete, but probably competition between organisms of the same and different species is more important than competition between one species and another species. Otherwise, they do very little. Above the level of the species, genera and higher taxa never do anything. Clusters of related clones in this respect are the same as genera. They don't do anything either." Eldredge agrees and notes the irony in the fact that the very people who argued most forcefully for the real existence of species went on to deny them any significant role in the evolutionary process. Once species selection is properly understood, Eldredge (1985, p. 160) is forced to conclude that species result from the evolutionary process but do not function in it: "Species, then, do exist. They are real. They have beginnings, histories, and endings. They are not merely morphological abstractions, classes, or at best classlike entities. Species are profoundly real in a genealogical sense, arising as they do as a straightforward effect of sexual reproduction. Yet they play no direct, special role in the economy of nature."

The point of the preceding is to jar those who are complacently satisfied with traditional, commonsense conceptions. Anyone who thinks that the preceding pages are excessive has never urged a nonstandard view on an intellectual community. The most common response is furious frustration. The world *must* be the way that it *seems*. The certainty with which such observations are proclaimed is historically quite contingent. No longer do ordinary people stamp their feet in frustration as they insist that the Earth must be in the middle of the universe or that space cannot possibly be curved. Many do continue to insist that species cannot possibly evolve. But such responses are not limited to ordinary people. Scientists (not to mention philosophers) are just as prone to such responses when their current commonsense perceptions are challenged. It is one thing to claim that over great stretches of space and time "straight" is "curved." Such expanses are not part of common sense, but organisms and species are. Hence, any attempt to alter how we view these entities is as threatening as any conceptual alteration can be.

As if treating genes, organisms, and species as the same sort of thing were not sufficiently counterintuitive, I have argued elsewhere that stratifying the organizational hierarchy in biology into genes, organisms, and species is "unnatural" (Hull 1980). I am not objecting to a hierarchical view of evolution (Arnold and Fristrup 1981; Eldredge 1985; Eldredge and Salthe 1984; Plotkin and Odling-Smee 1981). To the contrary, I think that evolution must be viewed hierarchically. Instead I am arguing that the traditional gene/organism/species hierarchy is seriously misleading. Common sense notwithstanding, it is "unnatural," and it is unnatural in just those respects that make the evolution of behavioral and conceptual evolution look so nonstandard. Objecting to *one* particular hierarchical ordering is not quite equivalent to objecting to all such orderings.

For me, a way of dividing the world is "natural" if it produces entities that function as such in general processes. If "genes" are a natural level of organization distinct from organisms and species, then there must be some function that genes and only genes perform in some natural process. As genes are commonly conceived, I think that there are no general processes in which genes and only genes function. The same can be said for organisms and species. There is nothing that all and only organisms do; nothing that all and only species do. To state the obvious: when I claim that there is no function performed by all and only genes, or organisms, or species, it does not follow that these entities perform no functions in any natural process; only that these are not natural subdivisions. Given a particular function, most genes and some organisms might perform it; given another function, most genes and organisms plus some species might perform it; and so on.

At one time, the division of plants into trees, bushes, and plants (herbs) seemed quite natural. Biologists now find it to be of no significance whatsoever. Currently, most people, including most biologists, find the distinction between genes, organisms, and species to be just as natural. It is not. One reason why evolutionary biologists have been unable to discover universal regularities in the evolutionary process is that they are not comparing like with like. They are dividing up the organizational hierarchy inappropriately. The appropriate levels are not genes, organisms, and species as they are traditionally conceived, but replicators, interactors, and lineages. Evolution needs to be viewed as a hierarchical process. The issue is the character of this hierarchy. My claim is that the regularities that elude characterization in terms of genes, organisms, and species can be captured if natural phenomena are subdivided differently: into replicators, interactors, and lineages. If not, this reworking of biological common sense serves no purpose whatsoever.

Previously I have set out the distinction between replicators, interactors, and lineages with respect to biological evolution (Hull 1980) and have shown how they might be extended to social learning and conceptual change (Hull 1982), but my emphasis in those papers is on replication. Here I expand on the notion of interaction with respect to both biological and conceptual evolution. Neither biological evolution nor social learning can be understood adequately entirely in terms of replication. The process that I term *interaction* plays too central a role to be omitted. The sort of social learning with which I am most concerned occurs in science. The account I set out is intended to apply to conceptual change in general, but the sort of conceptual change I emphasize is the sort that takes place in science.

REPLICATORS AND INTERACTORS

In his classic work, Williams (1966) redefined *gene* so that the extent of the genetic material that counts as a single gene depends upon the effects of selection. Dawkins (1976, p. 69) introduced the more general notion of a *replicator* to take the place of Williams's *gene*. Replicators include genetic replicators but "do not exclude any entity in the universe which qualifies under the criteria listed." According to Dawkins, replicators are those entities that pass on their structure intact through successive replications. Identity of structure is not good enough for selection processes. Identity by descent is required. However, identity seems a bit stringent for the individuation of replicators. Mutations with varying degrees of effect do occur. Allowing variations that have minimal effect on the functioning of a stretch of the genetic material to count as replicates of the "same" gene would not do excessive damage to the spirit of Williams and Dawkins's notion. Abandoning the requirement of descent would.

In his early writings, Dawkins (1976) emphasized replication so strongly that many of his readers interpreted him as arguing that replication is not just necessary for selection but also sufficient. In the interim Dawkins has "clarified" his position or, as his critics claim, "changed" it (Sober 1984). In any case, according to Dawkins's (1982a,b) current views, replication is necessary but not sufficient for selection. A second process, which I term *interaction*, is also necessary (Hull 1980). Interactors are those entities that interact as cohesive wholes with their environments in such a way as to make replication differential. Thus, selection can be characterized generally as any process in which differential extinction and proliferation of interactors causes the differential perpetuation of the replicators that produced

22

them. Vrba (1984, p. 319) phrases this same definition of selection as follows: "Selection is that interaction between heritable, emergent character variation and the environment which causes differences in birth and/or death rates among variant individuals within a higher individual."

The most important feature of the preceding definitions of selection is that selection involves more than just correlations. As Sober (1981, 1984) and Brandon (1982) emphasize, selection is a *causal* process. In my terminology, replicators are causally related to interactors, and the survival of these interactors is causally responsible for the differential perpetuation of replicators.[1] Brandon and Burian (1984) and G. C. Williams (personal correspondence) have complained that my definition of selection mistakenly includes drift as a form of selection. However, when the notion of *interactor* included in this characterization of selection is unpacked, drift is excluded. An entity counts as an interactor only if it is functioning as one in the process in question. It is not enough that in past interactions it functioned as an interactor. Thus, if changes in replicator frequencies are not being caused by the interactions between the relevant interactors and their environments, then these changes are not the result of selection. In instances of drift, there may be genes and organisms, but there are no interactors, only replicators.

Like Dawkins's notion of replicator, *interactor* is defined with sufficient generality that it is not necessarily limited to one common-sense level of organization. Certainly organisms are paradigm interactors, but entities at other levels of the traditional organizational hierarchy can also function as interactors. Genes, chromosomes, and gametes interact with their environments just as surely as do organisms, and these interactions can influence replication frequencies. Entities more inclusive than organisms can also function as interactors – e.g., colonies, hives, and other forms of kinship groups. If the traditional organizational hierarchy is retained, then both replication and interaction wander from level to level. The obvious solution to this state of affairs is to replace the traditional organizational hierarchy with a hierarchy whose levels are delimited in terms of the evolutionary process itself.

The distinction between replication and interaction is important because it helps to disambiguate the phrase "unit of selection." When gene selectionists say that genes are the primary units of selection, they mean that genes are the primary units of replication. They do not mean to assert that they are the only or even primary units of interaction. For example, Williams (1966) emphasizes the role of genes in replication without proposing that evolution is nothing but changes in gene frequencies. Organisms play as large a role in his discussion as in the writings of his critics, in some cases more so. Conversely, when organism selectionists insist that organisms are the primary units of selection, they mean that organisms are the primary focus of interaction, not of replication. Similar remarks hold with respect to group selection. When Wilson (1975) insists that colonies can function as units of selection, he does not mean that they are replicators; he means that they form higher-level interactors. Some species selectionists seem to maintain that species can sometimes function as replicators, others that they might well function as interactors (Dawkins 1982a; Eldredge 1985; Williams 1985).

Although distinguishing replicators from interactors helps to clarify the disagreements between advocates of various sorts of selection, it does not eliminate them completely. Proponents of group selection insist that close kin form groups and that these groups function as interactors in the evolutionary process. Organism selectionists counter that everything that needs to be said about kin selection can be said in terms of the inclusive fitness of individual organisms. The difference is between kin-group selection and kin selection. Gene selectionists dismiss inclusive fitness as a sop that Hamilton (1964a,b) threw to organism selectionists. Gene selectionists acknowledge that both replication and interaction function in the evolutionary process, but they maintain that evolutionary theory can, at bottom, be couched entirely in terms of replication, and that any causal processes that do not eventuate in changes in replicator frequencies (usually gene frequencies) simply do not matter. Opponents of genic selectionism, including Wimsatt (1980) and Sober and Lewontin (1982), admit that reference to changes in gene frequencies is adequate for the "bookkeeping" aspect of selection but insist that the bookkeeping aspect by itself leaves out too much of the causal story. They want to expand the axioms of evolutionary theory to include reference to the causal relations responsible for evolution's taking place the way that it does. Williams (1985, p. 2) is content with evolutionary theories limited to the bookkeeping aspect and finds the criticisms of such theories by Wimsatt (1980) and Sober and Lewontin (1982) to be based on "unrealistic expectation." Genic selectionists also seem to fear that including explicit reference to the causal interface between interactors and their environments will complicate evolutionary theory prohibitively, possibly even challenging the Weismann barrier.

Strangely enough, the most vocal defenders of the necessary role of organisms (or, more generally, interactors) in the evolutionary process – in opposition to what they see as an overemphasis on genes (or, more generally, replicators) – are also among those who are most skeptical about evolutionary "just so stories" (e.g., Gould and Lewontin 1979), while several of the strongest advocates of genic selectionism see nothing wrong with evolutionary scenarios (Dawkins 1982b; Ridley 1983; Williams 1985). Defenders of the role of organisms in evolution warn that many of the characteristics that evolutionary biologists claim are adaptations might well be nothing but effects. In addition, the ease with which adaptationist scenarios can be constructed to explain particular adaptations casts considerable doubt on the entire program. Hence, they conclude that we cannot leave organisms and their adaptations out of the evolutionary story, but we must include reference to them only with great care. Defenders of the adequacy of genes in setting out the basic axioms of evolutionary theory do not think that discerning organismic adaptations is all that problematic. Although organisms and their adaptations do not play a central role in the fundamental axioms of evolutionary theory, secondary reference to them is nevertheless scientifically reputable. At the risk of putting too fine a point on the dispute, Williams (1985, pp. 2, 15) thinks that his critics are too optimistic about the potentialities of evolutionary theory and too pessimistic about the legitimacy of adaptive scenarios.

I find myself in partial agreement with both sides of this dispute. I think that any adequate theory of evolution must include reference to the interactor-environment interface (Odling-Smee and Plotkin 1984), but that the inclusion of such reference need not complicate evolutionary theory any more than including reference to replication does. It is certainly true that interactions are as varied as the myriad causal situations that give rise to the incredible array of adaptations that makes the study of biology so endlessly fascinating, but the information contained in the genetic makeup of organisms for these adaptations is just as multifarious. The introduction of either sort of complexity into the general characterization of the evolutionary process would be lethal, but no such introduction is necessary. In both cases, all that must actually be included in formal statements of evolutionary theory are the general characteristics of replicators and interactors and how they are interrelated. Only when this general theory is applied to particular cases is the actual informational content of the replicators and the causal situations that produced the particular adaptations relevant. Scientific theories are general. Their applications are contingent and often idiosyncratic.[2]

With respect to applications, I see no reason to shy away from claiming that a particular characteristic arose as an adaptation to a putative past environment, even though such claims may often be false. In most cases, little rides on the correctness of particular adaptationist scenarios. Showing that processes other than replication and interaction are actually responsible for biological evolution would be of prime importance. It would bring into question our basic conception of the evolutionary process. Detailing difficulties in applying evolutionary theory is of secondary importance. All scientific theories are difficult to apply. Inferences to particular cases must be possible if evolutionary theory is to be testable, but such testing need not be easy or automatic. Critics of evolutionary theory are not content with its being falsifiable. They insist that it must be easily falsifiable, when no scientific theory is easily falsifiable.

However, adaptationist scenarios are so fascinating that they often seduce biologists into ignoring even more fundamental aspects of the evolutionary process. Also, there is a tendency to think that adaptationist scenarios have greater warrant than they actually have. For most species, such misplaced confidence is unlikely to do much harm. However, similar mistakes in the context of the human species can do considerable damage. We may be innately territorial or sexually dimorphic in socially relevant ways. We *may* be, but the substantiation available for such claims is not all that impressive. Social policies based on such shaky ground are likely to be misconceived and the results deleterious. But to repeat my general point: The general notion of adaptation is central to a selectionist view of evolution. Questions about which particular structures arose as adaptations to which particular environmental changes are relevant only to the testing of selectionist versions of evolutionary theory.

Although Dawkins has come to accept the distinction between replicators and interactors, he prefers a somewhat different terminology: "My main concern has been to emphasize that, whatever the outcome of the debate about organism versus group as *vehicle*, neither the organism nor the group is a *replicator*. Controversy may exist about rival candidates for replicators and about rival candidates for vehicles, but there should be no controversy over replicators *versus* vehicles.

Replicator survival and vehicle selection are two aspects of the same process" (Dawkins 1982a, p. 60).

Once the distinction between replicators (on the one hand) and vehicles or interactors (on the other hand) is made, the issues that divide present-day evolutionary biologists can be stated more clearly. Of course, this distinction does not decide these issues. Slight differences in how the distinction is drawn can influence the resulting resolutions. As Kawata (1987) points out, Dawkins's vehicles and my interactors are not precisely equivalent concepts. According to Dawkins, genes are replicators, not vehicles. They ride around in vehicles, directing their behavior. On my account, genes are both replicators and interactors. If genes are anything, they are entities that interact with their environments in such a way as to bias their own replication. In one place, Dawkins (1982a, p. 45) notes that the wings of birds are for flying and then asks, "What are DNA molecules for?" He answers that "DNA is not 'for' anything. If we wish to speak teleologically, all adaptations are for the preservation of DNA; DNA itself just *is*." Yes and no. Organisms are characterized by adaptations. So are molecules of DNA. They are extremely well adapted to replicate. The major effect of this replication is, as Dawkins insists, the preservation of the structure of DNA. Once it is recognized that one and the same entity can function both as a replicator and as an interactor, the image of genes riding around in vehicles becomes less persuasive.

Sober (1984, pp. 253–5) also complains that Dawkins defines his terms in ways to guarantee that organisms cannot possible function as replicators. One reason that Dawkins has for rejecting organisms as replicators is meiosis. At meiosis, parental genomes are dismembered; then new genomes are reassembled at fertilization. If retention of structure largely (or totally) intact is necessary for replicators, then only small sections of the genome can function as replicators in cases of sexual reproduction in genetically heterogeneous populations. Both restrictions should be noted. Dawkins's central argument for genes being the only replicators applies only to organisms when they reproduce sexually. If genes are the only replicators in cases of asexual reproduction, he needs an additional argument. His central argument also does not apply to sexual reproduction in genetically homogeneous populations. Although crossover can occur in such populations, it makes no difference to the structure of the resulting genomes. However, as Williams (1985, p. 5) notes, even in such cases, phenotypes "can play no role in bookkeeping because, even in a clone, the successive generations of phenotypes may be markedly different because of environmental variables that affect development."

The best example of an organism functioning as a replicator is the direct transmission of a phenotypic change to successive generations through fission. For example, if a portion of the cortex of a paramecium is surgically removed and reinserted with the cilia facing the opposite direction, this phenotypic changes is transmitted to subsequent generations. In light of this example, Dawkins (1982b, p. 177) responds as follows:

> If, on the other hand, we look at underlying replicators, in this case perhaps the basal bodies of cilia, the phenomenon falls under the general heading of replicator propagation. Given that macromolecular structures in the cortex are true replicators, surgically rotating a portion of cortex is analogous to cutting out a portion of chromosome, inverting it, and putting it back. Naturally the inversion is inherited, because it is part of the germ-line. It appears that elements of the cortex of *Paramecium* have a germ-line of their own, although a particularly remarkable one in that the information transmitted does not seem to be encoded in nucleic acid.

When Dawkins (1976) introduced the term *replicator*, he intended it to be more general than *gene*. Although genes, as they are currently understood, may not be limited to the nucleus, they are limited to nucleic acids – DNA and RNA. The basal bodies of cilia do not count as genes, but they might well count as replicators and form a "germ-line" of their own. Even so, in the paramecium example, the inverted cortex is not being transmitted indirectly via the basal bodies but directly. Obviously something must be going on at lower levels of analysis when a paramecium splits down the middle to form two new organisms, but I fail to see why this fact counts against treating organisms in this situation as replicators. After all, even though entities less inclusive than genes are involved in genetic replication, it does not follow that genes are not replicators. Standards should not be invoked for organisms more stringent than those applied to genes. Organisms behave in ways that make them candidates for replicators seldom enough without ruling them out by definition. As it turns out, in the most common situation in which one might want to view organisms as replicators – asexual reproduction via fission – it makes no difference. In asexual reproduction, usually the entire genome functions as a single replicator, and there is a one-to-one correlation between genomes and phenomes. Hence, the numbers will always turn out to be the same.

A second reason that organisms are not very good candidates for replicators concerns the different senses in which genes and organisms can be said to "contain information" in their structure. Genes do not code in a one-to-one fashion for phenotypic traits. Given a particular genome, numerous alternative phenomes are possible depending on differences in the environment. Given any one gene, numerous different alternative traits are possible depending on differences in the environment as well as elsewhere in the genome. The net effect is that both individual genes and entire genomes code for reaction norms, not for specific traits or phenomes. In this sense, the information in a genome is largely "potential." In any one instance of translation, these reaction norms are narrowed to one eventuality, to a single phenome. All other possibilities, equally "programmed" into the genetic material, are not realized. To use Wimsatt's felicitous terminology, each genotoken gets to produce a single phenotoken. The net effect is the loss of nearly all the potential information in the genome. The only information that an organism as a replicator can pass on is the information realized in its structure. (For an exhaustive treatment of the role of "information" in evolution, see Brooks and Wiley 1986.)

Both Dawkins (1982b) and I (Hull 1976, 1978a) have been concerned to break the hold that a fairly narrow conception of organisms has on the minds of many evolutionary biologists, but toward different ends. According to Dawkins, nests and mating calls are as much a part of a bird's phenotype as are its beak and webbed feet. In response to Gould's (1977) claim that selection cannot see genes and pick among them directly but must use bodies as an intermediary, Dawkins (1982a, p. 58) retorts: "Well, it must use *phenotypic effects* as intermediaries, but do they have to be bodies? Do they have to be discrete vehicles at all?" In answering no to both questions, Dawkins plays down the role in evolution of organisms as discrete bodies or even as vehicles. I have argued at some length that organisms are not as discrete, unitary, and well organized as we tend to assume in order to urge a role for entities more inclusive than single organisms as interactors, not in order to question the role of organisms as paradigmatic interactors. Dawkins (1982b) and Williams (1985) argue that organisms can never function as replicators in the evolutionary process. Although I am not willing to go this far, I agree that when organisms do function as replicators, the effects of organismal replication are likely not to be extensive.

The point of the preceding discussion has been to show why traditional conceptions are not adequate for biological evolution strictly construed. If simplistic notions of genes, organisms, and species are not adequate for ordinary biological evolution, then they are surely not adequate for construing social learning as a selection process. But before turning to this topic, I need to present at least one particular example of the improved understanding that my revised conceptual apparatus brings to biological phenomena. One of the major topics in recent literature in population biology has been explanations of the prevalence of sexual reproduction. The problem can be stated quite simply. If the name of the game in biological evolution is to pass on one's genes, then sexual reproduction is a very inefficient way of accomplishing this end, because sexual reproduction has a 50 percent cost. At any locus where the male and the female differ, each has only a 50-50 chance of passing on its alleles rather than those of its mate. Conversion from sexual reproduction to parthenogenesis would double the contribution of a female to future generations. So, though in theory sexual reproduction should be quite rare, in fact it is "almost universal" (Maynard Smith 1971, p. 165).

The problem is so acute that evolutionary biologists who are strongly inclined to dismiss group selection in other contexts are forced back on this mechanism for the evolution of sex. For example, Williams (1971, p. 161) concludes: "Sexual reproduction must stand as a powerful argument in favor of group selection, unless someone can come up with a plausible theory as to how it could be favored in individual selection. And if group selection can produce the machinery of sexual reproduction, it ought to be able to do many other things as well."

Among the many things that group selection has been introduced to explain is the evolution of sociality and, with it, the rise in importance of social learning. Hence, sex and society are intimately connected in theorizing about the evolutionary process. But in the preceding discussion, the most important premise gets slipped in when no one is noticing, i.e., that sexual forms of reproduction are prevalent. As my earlier discussion indicates, sexual reproduction is a relatively recent innovation. For three-quarters of the existence of life on Earth, the sole form of reproduction was asexual. If one looks at every measure save one, it is still extremely common. If one looks at number of organisms, amount of energy transduced, biomass, etc., asexual reproduction remains extremely prevalent. Only if one compares numbers of species do sexual forms of reproduction turn out to be "nearly universal." But asexual organisms do not form species of the sort that exist among sexual organisms. To be sure, systematists group all organisms in species (taxospecies); however, as far as real groupings in nature are concerned, asexual organisms do not form *genealogical* units of the sort formed by sexual organisms.

The difference between asexual and sexual reproduction is fundamental. As Maynard Smith (1971, p. 163) notes, "At the cellular level, sex is the opposite of reproduction; in reproduction one cell divides into two whereas it is the essence of the sexual process that two cells should fuse to form one." In fact, the differences are so fundamental that many authors argue that the same term should not be applied to both. Either sexual or asexual reproduction is not really "reproduction." Hence, one solution to the problem of the prevalence of sex when meiosis exacts a 50 percent cost is that it is not prevalent. It is as rare as it should be given its cost. The reason why it took so long for sex to evolve is that it is advantageous in only very special circumstances. In fact, it took only a little over a billion years for the first living creatures to evolve. It took almost 3 billion years more for sexuality to make an appearance. If the time it takes for something to evolve is any measure of its evolutionary advantage, sexual reproduction may not be all that advantageous. Hence, from this perspective, Williams's (1985, p. 103) explanation of sexuality in vertebrates begins to sound more plausible. According to Williams, sexual reproduction in derived low-fecundity organisms such as vertebrates is "a maladaptive feature, dating from a piscine or even protochordate ancestor, for which they lack the preadaptations for ridding themselves."

The usual response to the preceding observations is that something has gone wrong. Sexual reproduction evolved quite early and has been widespread throughout the history of life on Earth. After all, forms of parasexual reproduction exist among extant blue-green algae. There is no reason to assume that such forms of gene exchange were any less prevalent in the past. In the first place, mere gene exchange does not pose the same problem as meiosis. The issue is the cost of meiosis. And by all indications gene exchange among prokaryotes is extremely rare, ranging from one cell in 240,000 replications to one in 20 million. If such rare occurrences of gene exchange are sufficient to label an entire higher taxon "sexual," then Jackson et al. (1986) are just as warranted in labeling an entire group "clonal" just because a few forms exhibit clonality. One need not argue that sexual reproduction evolved quite early and is nearly universal in order to recognize it as an important innovation in biological evolution. Even though it is a relatively recent innovation and still quite rare, it served as an "evolutionary trigger" to give rise to species and, through them, to much of the diversity of life that we see all around us.

On the view that I am urging, replicators should be compared with replicators, interactors with interactors, and lineages with lineages. When one makes such comparisons, the results are quite different than when one compares genes with genes, organisms with organisms, or species with species. For one thing, sexual reproduction becomes "rare." Although their terminology is different, those authors who have looked at clonal organisms have been forced to make very similar distinctions. When the authors in Jackson et al. (1986) look at evolution in clonal organisms, they are forced to distinguish between ramets and genets. They compare ramets with ramets and genets with genets in estimating such things as fitness and the speed of evolutionary change. The effects of this change in perspective are dramatic in biological evolution. They should be no less pervasive when one turns one's attention to social learning as a selection process.

CONCEPTUAL EVOLUTION: REPLICATION

Dawkins introduced the notions of replicator and vehicle because of their generality and because of the common associations of such terms as *gene* and *organism*. However, *replicator* and *vehicle* also have their connotations. As far as I can see, the connotations of the term *replicator* are entirely appropriate whereas those of *vehicle* are not. Vehicles are the sort of thing that agents ride around in. More than this, the agents are in control. The agents steer and the vehicles follow dumbly. The picture that Dawkins's terminology elicits is that of genes controlling helpless and hapless organisms. Although Dawkins explicitly assigns an evolutionary role to both replicators and vehicles, his terminology is likely to mislead one into treating vehicles as passive tools in the hands of all-powerful replicators. As Sober (1984, p. 255) repeatedly emphasizes, "The units of selection controversy began as a question about causation." For this reason, I prefer *interactor* to *vehicle* (see also Williams 1985).

Dawkins intends *replicator* to apply to any entity that happens to possess the appropriate characteristics. In biological evolution, he insists that only genes function as replicators (the paramecium example notwithstanding). However, Dawkins (1976, p. 68) recognizes that in other sorts of selection processes other entities might function as replicators – for example, in "cultural analogues of evolution." He terms the cultural analogues of genes *memes* (see also Semon 1904). According to Dawkins, genes and memes are equally replicators. If memes are to function as replicators, then they must have structure and be able to pass on this structure through successive replications. If conceptual change is to occur by means of selection processes, memes cannot exist in some other "world" (Popper 1972) but must exist in the material world – in brains, computers, books, etc. A second reason for preferring *interactor* to *vehicle* is that the father of evolutionary epistemology, Donald Campbell (1979), uses *vehicle* to refer to replication in both biological and conceptual evolution. Genes are the vehicles that transmit the information in biological evolution, whereas everything from stone tablets and papyrus to magnetic tapes and electronic chips can serve as the physical vehicles in conceptual evolution. Using *vehicle* to refer both to interactors and to the physical basis of replication begs for misunderstanding, and misunderstanding comes along easily enough on its own. One need not beg for it.

Thus far, the burgeoning literature on conceptual change as a selection process has concentrated primarily on conceptual replication and how it differs (or does not differ) from replication in biological evolution. The most common alleged disanalogies between the two processes are that conceptual evolution is Lamarckian whereas biological evolution is not, that conceptual evolution is not biparental the way that biological evolution is, that cross-lineage borrowing is common in conceptual evolution but rare or nonexistent in biological evolution, and that conceptual evolution can be and often is insightful and intentional whereas biological evolution is blind and mechanical. Elsewhere I have argued that these alleged disanalogies are exaggerated, and that they stem from the failure to distinguish adequately between gene-based biological evolution and meme-based conceptual change (Hull 1982).

Though proponents and opponents of treating conceptual change as a selection process have often claimed that conceptual evolution is somehow "Lamarckian" no one has explained at much length what this term means in connection with conceptual change. In biological evolution, inheritance counts as "Lamarckian" if adaptive changes in the phenotype of an organism were transmitted to the genetic material and thereafter inherited by the organism's progeny. Acquired characteris-

33

tics must be *inherited*, not just transmitted. The above example of alterations in the cortex of a paramecium is not an example of Lamarckian inheritance because the genetic material is bypassed. Social learning would be literally Lamarckian if the knowledge that an organism acquired about its environment somehow came to be encoded in its genetic material and thereafter was inherited by its progeny. As far as I know, none of the advocates of an evolutionary analysis of conceptual change view social learning in such a literal fashion. The whole point of social learning is that information is transmitted independently of genes. If social learning is Lamarckian, it must be so only in a metaphorical sense of this term. Such conceptual entities as memes must be substituted for genes, but it should be noted that memes are analogous to genes, not to characteristics. Hence, their transmission does not count as an instance of the inheritance of acquired characteristics precisely because they are not the analogues of characteristics. In sum, on a literal interpretation, social learning is not an example of the inheritances of acquired characteristics because inheritance is not involved (just transmission). On a metaphorical interpretation, social learning does not count as an instance of the inheritance of acquired characteristics because the things being passed on are analogues of genes, not of characteristics. Social learning is, if anything, an instance of the inheritance of acquired memes. One organism can certainly give another fleas, but this is hardly an instance of the inheritance of acquired characteristics. Social learning is to some extent "guided" (Boyd and Richerson 1985), but to call it on that account "Lamarckian" is to use this term in its most caricatured form, as if giraffes got such long necks by striving to reach the leaves at the tops of trees.

In this connection, commentators often state that biological evolution is always "vertical" whereas conceptual evolution is likely to be "horizontal." By this they mean that the transmission of characteristics in biological evolution is always from parent to offspring (i.e., inheritance). Characteristics always follow genes. In point of fact, biological evolution is not always vertical, even when characteristics follow genes. For example, it is horizontal when bacteria, paramecia, etc. exchange genetic material. Horizontal transmission can even be cross-lineage, as when viruses pick up genes from an organism belonging to one species and transmit them to an organism belonging to a different species. In conceptual contexts, parents can instruct their offspring, but they can also teach things to their elders, to others of their own biological generation, or to younger organisms to which they are not closely related. From the perspective of gene lineages, considerable cross-lineage borrowing occurs in conceptual evolution, but all this shows is that the relevant lineages for conceptual evolution are not gene lineages. The transmission of memes is what determines conceptual lineages. Hence, by definition, if a significant amount of cross-lineage borrowing is taking place between two conceptual lineages, these are not two conceptual lineages but one. The situation is exactly analogous to the situation in biological evolution. If a significant amount of gene exchange is taking place between two putative lineages, these lineages count not as two lineages but one (Hull 1982, 1984a, 1985a).

Sometimes conceptual change is "biparental" – that is, ideas are obtained from two different sources and combined – but quite obviously information can be transmitted from a single source to another or combined from several sources. If the transmission of genes were actually always biparental, this would be an important difference between biological and conceptual evolution, but of course it is not. Both asexual reproduction and polyploidy are common. In general, those who oppose treating conceptual change as evolutionary reason from an extremely impoverished view of biological evolution to the context of conceptual evolution. Their view of biological evolution is so narrow that most biological evolution does not fit.

In this same connection, commentators on an evolutionary analysis of conceptual change are nearly unanimous in noting that conceptual change can occur much more rapidly than biological evolution (for an exception see Boyd and Richerson 1985). For example, under the most extreme selection pressures, a mutation that arose in the time of Julius Caesar would only now be becoming widely distributed in the human species. In this same time interval, conceptual systems have undergone great changes several times over. But the preceding contrast depends on taking calendar time as the appropriate time frame for both biological and conceptual evolution, when it is adequate for neither. Biological evolution is phylogenetic; it occurs only through a succession of biological generations. Individual learning is ontogenetic. It takes place within the confines of a single biological generation. In this respect it is like the immune system. Social learning is both ontogenetic and phylogenetic. It can occur both within and between biological generations. However, neither calendar time nor biological generations is the univocal time frame appropriate for either biological or conceptual selection processes.

With respect to calendar time, bacteria reproduce much more quickly than elephants; from the perspective of generations, they reproduce at the same speed. One reason why claims about molecular clocks caused such consternation among evolutionary biologists is that they were supposed to be constant with respect to calendar time, regardless of the generation time of the organisms in which these changes were occurring. With respect to the evolutionary process, a variety of time frames are relevant. For example, for mutation, cell cycle time is more appropriate than the generation time of the entire organism (Lewin 1985); however, for the evolutionary process as such, calendar time enters in only with respect to ecological interactions. For example, because of differences in generation time, new strains of bacteria and viruses pose dangers for organisms with slower generation times. They themselves cannot evolve fast enough to keep up with the bacteria and viruses, but their immune systems can. As a result of the preceding considerations, the appropriate time frame for replication in conceptual evolution is generational. Each time a meme is replicated, that is a generation. Thus, in the course of his biological lifetime, a geometry teacher may replicate the Pythagorean theorem hundreds of times. From the perspective of physical time, conceptual generations are much shorter than certain biological generations and longer than others; but from the perspective of generations *per se*, biological and conceptual evolution take place at the same speed – by definition.

The only frequently alleged difference between biological and conceptual evolution that does not arise from a straightforward misunderstanding concerns intentionality. Intentionality certainly plays a role in biological evolution. Both human and nonhuman organisms strive to elude predators, find mates, etc. However, a small number of the organisms belonging to the human species are aware that species evolve. As a result, they are in a position to influence that evolution consciously. Members of all species influence the evolution of their own and other species *un*intentionally, but the few people who acknowledge the existence of biological evolution and understand it sufficiently are in a position to direct it intentionally. We already do so in the case of domesticated plants and animals. Most of the changes that we have wrought in these creatures have been unintentional, but some have been consciously brought about. In the past, we have had to wait around until a particular variation happened to crop up. We are now in a position to introduce specific variations and to select the resulting variants. It would seem that we have always been in this position in cases of conceptual change. For instance, scientists often strive to solve problems and in doing so intentionally direct the course of conceptual evolution. In conceptual evolution both the introduction of variations and their selection can be done consciously toward certain ends.

Whether or not intentionality presents a significant disanalogy between biological and conceptual evolution depends upon how we distinguish between the two. Two criteria have been suggested: the sort of entity that functions as the relevant replicators (genes versus memes) and the source of new variants and/or their subsequent selection (intentional or not). Given these two criteria, four combinations are possible. Two combinations pose no problems. Most biological change is gene-based and nonintentional. Neither the introduction of new variants nor their selection is in any sense intentional. Some conceptual change (probably not much) is meme-based and intentional. A conscious agent either generated the conceptual variant intentionally, or subsequently selected this variant, or both. But the other two combinations raise some difficulties. Some change is gene-based and intentional – selective breeding. The things being changed are genes, and the traits that are being selected are being transmitted via genes. However, the agent involved is conscious of what he or she is doing and is doing it intentionally. In Darwin's day the presence of a conscious agent in artificial selection and the absence of such an agent in natural selection was considered extremely important. In reasoning from artificial selection to natural selection, Darwin took himself to be reasoning by analogy. Just as breeders could select wisely, so could nature (Young 1971; Ruse 1975; Waters 1986). However, today artificial selection is considered to be a special case of natural selection and part of the legitimate subject matter of evolutionary biology – the presence of an intentional agent notwithstanding (Rosenberg 1985, p. 171).

The final combination is unintentional meme-based change. If Freud is right, understanding, inference, conscious choice, and the like play much less of a role in human behavior than his more rationalistic contemporaries thought. Although I am hardly a fan of Freud, I have a fairly skeptical attitude toward the role of these factors in human affairs. The rule that human beings seem to follow is to engage the brain only when all else fails – and usually not even then. However, the relevant issue is not the frequency of the relevant behavior but its classification as biological or conceptual. If the presence of intentionality is the crucial difference between biological and conceptual evolution, then artificial selection belongs in the province of conceptual evolution and all the unintentional conceptual changes produced by humankind belong to neither. I am not sure what choices the critics of an evolutionary analysis of conceptual change are likely to make in these matters. However, further discussion requires the introduction of the second aspect of selection processes, interaction. (For a more extensive discussion of the place of intentionality in nature, see Searle 1984.)

CONCEPTUAL EVOLUTION: INTERACTION

In biological evolution, entities at numerous levels of organization interact with their respective environments as cohesive wholes in such a way that replication is differential. Some sperm can swim faster than others, some antibodies are more effective than others, some kidneys are better able to eliminate wastes, some organisms can withstand dessication for longer periods of time, some beehives can keep their internal temperature more constant than others, and sexual reproduction may have arisen as a species-level adaptation to increase speciation rates (Lewontin 1970). Any characterization of biological evolution that leaves out reference to interactors and their adaptations is leaving out half the causal story. The same observation holds for conceptual change. If the notion of conceptual replication makes sense, the task of identifying conceptual interactors remains. To put the issue in more restricted terms: analogues to the genome-phenome distinction must be specified in conceptual evolution. In conceptual change, memes physically embody information in their structure. This structure is differentially perpetuated. But what is responsible for certain information proliferating while other information is lost?

One sure sign in biological contexts that autocatalysis (the transmission of information in replication sequences) is being replaced by heterocatalysis (the translation of information contained in the structure of the replicators) is a precipitous loss of "potential" information. In sexual reproduction, each genotype is almost always instantiated by a single genotoken, and this single genotoken usually gets to produce only a single phenotoken. Hence, in such circumstances, each genotype is selected via a single phenotoken. In cases of cloning, particular genotypes are represented by several genotokens. Hence, they can be tested by means of several phenotokens. But even in such cases, numerous alternative representations are never realized. Biological evolution seems "unfair" on a host of counts. One of them is that neither single

38

genes nor entire genomes ever get to show their "real stuff." They succeed or fail in replicating themselves, depending on a relatively small number of actual exemplifications of all possible exemplifications permitted by the information they contain.

The same is true of conceptual replicators. Natural languages serve many functions. One of them is communication. Another is description, and part of what is communicated are these descriptions. Communication is the analogue to replication, whereas the testing of descriptive statements is the analogue to interaction – the translation of the information contained in a descriptive statement in such a way that it can be tested. A single gene corresponds roughly to a single concept, an entire genome to a more inclusive conceptual entity such as a scientific theory. Just as single genes never confront their environments in isolation, single concepts are never tested in isolation.

Philosophers have argued at great length that the meaning of a theoretical term is never exhausted by the various operational "definitions" used to apply it. A particular experiment or observation bears on only one small part of the meaning of the theoretical claim. For descriptive statements, the analogue to the interactor-environment interface is testing. Any minimally sophisticated conceptual system implies a huge array of observational consequences. Only a very few are ever likely to be tested, but the system will be accepted, rejected, ignored, or modified on the basis of these few tests. Conceptual change is hardly less unfair than biological evolution. Sometimes just the right test is run in just the right way; at other times an unfortunate choice results in the rejection of a theory. Mendel's work on garden peas is an example of the first sort; his choice of a particular species of *Hieracium* to extend his theory is an instance of the second sort. Garden peas could not have been a better choice. They exemplify what has come to be known as Mendel's laws with admirable clarity. His second choice could not have been worse. Inheritance in *Hieracium* is near chaos.

Thus, the translation of a particular genome (genotoken) into a particular phenome that either does or does not survive to reproduce is equivalent to the testing of a particular descriptive statement (a conceptual token) in a particular context. Either it survives the test or it does not. In biological evolution, each genome is an instance of a genotype. Indirectly, then, the genotype has been tested, albeit inadequately. However, especially in cases of sexual reproduction, each genotype is instantiated only by a single genotoken. One reason for narrowing one's focus in studying evolution to small segments of the genetic material is that they are more likely to have numerous copies. The same genotype is likely to have numerous genotokens to be tested in a variety of contexts. Thus, some estimation of the relative "worth" of this genotype can be gathered. Similarly, conceptual systems of considerable scope are extremely complex. It is very unlikely that more than one scientist adheres to precisely the same global conceptual system. In fact, a given scientist is unlikely to retain allegiance to the same global conceptual system for very long. Scientists change their minds. Global systems are tested only in the form of "versions." What makes something a "version" is not just similarity in structure. Descent is also required. Theories are best interpreted as families of models (Giere 1984), but these "families" have a necessary genealogical dimension. That the comparison just outlined is appropriate is indicated by the massive loss of information in both contexts and the messiness at the relevant interfaces. Only one small aspect of a scientific theory can be tested in a particular experimental setup, and the results can always be accommodated in a host of ways (in part because in any test too many concessions must be made to experimental contingencies). There are no absolutely crucial experiments.

In my discussion of both biological and conceptual evolution I have emphasized the essential role of tokens ordered in lineages. The primary replicators in biological evolution are genotokens ordered into gene lineages. The primary replicators in conceptual evolution are conceptual tokens ordered in conceptual lineages. Is there no role for types – similar tokens regardless of descent? In biological evolution, there might well be. For example, albinism, eusociality, and photosynthesis apparently have each evolved numerous times. They are all tokens of the same type. From the point of view of phylogenetic descent, they are convergences – homoplasies rather than homologies. As such, their use in reconstructing phylogeny is likely to produce error. But there is more to evolutionary biology than phylogeny reconstruction. There is, for instance, the formulation of general statements concerning the evolutionary process, and one thing that is certain about the concepts incorporated in such general statements is that they must refer to types of phenomena. In this connection I do not think that either albinism or the ability to photosynthesize is a likely candidate for a type to function in general statements about the evolutionary process; eusociality and sexuality may be. If the concepts that function in statements of purported laws of nature are termed "natural kinds," then evolutionary biologists have not been tremendously successful in identifying natural kinds in the evolutionary process. One purpose of introducing such terms as *replicator, interactor,* and *lineage* is to specify class terms (types) more general than the traditional terms *gene, organism,* and *species.*

In the preceding discussion of the evolutionary process, terms such as *replicator* refer to types of entities. Anything anywhere that has the right characteristics counts as a replicator. It just so happens that included among these characteristics is temporal continuity. The entities themselves are historical entities; the type is not. However, when one moves to the level of conceptual evolution, *replicator* itself must be interpreted as a historical entity – a conceptual historical entity. As do all concepts, the term *replicator* has a history. Anyone who wants to understand the development of this concept must trace its history, and all the problems in distinguishing "homoplasies" and "homologies" arise. For example, Dawkins (1976) coined the term *meme* independent of Semon's (1904) earlier neologism. However, is there no role for type terms in our understanding of conceptual change? I think there is, just so long as one realizes that the instances of these type terms are themselves historical entities. To understand conceptual evolution, one must have a basic framework of conceptual historical entities. Periodically, a particular agent elaborates a set of conceptual entities in ways he or she takes to be genuinely general. These concepts will be evaluated as genuinely general (types with similar tokens), but in transmission this generality is lost once again. Only a few tokens actually get transmitted. The image that comes to mind is successive bursts of skyrockets. In each inflorescence, most of the rockets fizzle out; but a few explode into additional inflorescences, and so on (Hennig 1969, p. 43; Sneath and Sokal 1973, p. 321). Instead of treating the historically unrestricted types as constituting the general framework in which conceptual change is investigated, as is usually done, an evolutionary analysis takes a phylogenetic framework as basic; then conceptual types are periodically fitted into the interstices of this tree.

Numerous problems have been raised in connection with the testing of such conceptual systems as scientific theories that have nothing special to do with an evolutionary analysis as distinct from other analyses of conceptual change. However, one recurring problem that is particularly relevant concerns the social dimension of conceptual systems. Words do not confront the world in all their nakedness. Words do not mean anything. Instead, people mean things by the words that they use.

In many semantic theories, people drop out and are replaced by an abstract relation between word and object or statement and state of affairs, a relation that ignores all characteristics of the actual meaning situation save the proposed isomorphisms. Omitting reference to the interaction in conceptual change leaves out not only the testing part of conceptual change but also the tester – in cases of science, the scientist.

As I have noted, both gene selectionists and organism selectionists find replication adequate to handle the bookkeeping aspect of biological evolution. If there is a "bookkeeping" aspect of conceptual change, it is embodied in simple changes in meme frequencies. Internalist historians of science are frequently chastised for leaving too much out of their histories of science, but even the most internalist historians include references to scientists in their histories. Scientists are the ones who devise and evaluate scientific theories. The relevant weakness of internalist histories is not that they omit reference to scientists but that they omit what is commonly termed the "social context" of science. However, one reason why many philosophers of science – among them Collins (1975), Bloor (1976), and Barnes (1977) – feel uneasy about reference to "social context" is that they fear that it signals a relativist view of truth, and in many cases they are right. However, such references can also signal a relativist view of meaning. For example, Kitcher (1978) proposes to avoid some common semantic problems by postulating a community-based reference potential for each expression type. The reference potential of an expression type for a particular community is the "set of events such that production of tokens of that type by members of the community are normally instituted by an event in the associated set" (Kitcher 1978, p. 540).

Analyzing meaning in the context of communities of language users is certainly a step in the right direction for an evolutionary analysis of conceptual change, but several points must be emphasized. First, the communities must be defined by the appropriate relations, including such social relations as communicating with one another. If "reference potential" is to be of any use, communities cannot be defined in terms of their members' meaning the same things by the terms that they use. If communities are defined by the appropriate social relations, such as writes-to, reads-the-papers-of, and uses-the-work-of, one thing becomes clear: that plenty of conceptual heterogeneity exists in such communities (Hull 1984a, 1985a). Instead of being a weakness, such heterogeneity is a strength. If biological evolution is to occur by selection, variability is necessary – both intra- and interspecific variability. If conceptual evolution is to occur by means of selection, both intra- and intercommunity variability must exist, and it does. One of the chief strengths of Kuhn's (1970) analysis of scientific change is that he views it as a community-based activity. One of its chief weaknesses is that he thinks that all scientists belonging to the same scientific community share the same "paradigm." As Kuhn (1970, p. 176) puts his position, "A paradigm is what the members of a scientific community share, *and*, conversely, a scientific community consists of men who share a paradigm."

Whether Kuhn intends for his "paradigms" in the preceding statement to be entire disciplinary matrices or particular exemplars (his two primary uses of the term *paradigm*), his position simply will not do. Because Kuhn portrays communities as monolithic entities, the transition from one paradigm to another seems a highly problematic affair – so problematic that some of Kuhn's readers have interpreted him as claiming that it is arational. Actually, all Kuhn has claimed is that simplistic analyses of rationality cannot explain such transitions. A community-based notion of rationality is more appropriate (Sarkar 1982). However, once one acknowledges that considerable differences of opinion can exist within any socially defined community, the radical differences in kind between intragroup and intergroup communication disappear. There is often as much intragroup dissonance (incommensurability) as intergroup dissonance. To the extent that incommensurability is a genuine problem at all, it is as much of a problem within scientific communities as between them (Hull 1985a). In the life of a community, cooperation is more important than agreement. It is a contingent truth that the scientists who make up the small, ephemeral research groups that are so operative in science can disagree with one another without ceasing to cooperate.

The crucial feature of an evolutionary analysis of conceptual change is that conceptual tokens be ordered into conceptual lineages. Because human beings are among the chief vehicles for conceptual replicators, there will be a significant, though not perfect, correlation between communities and such lineages. In order to understand conceptual change, in science as elsewhere, both social groups (such as the Darwinians) and conceptual systems (such as Darwinism) must be interpreted as historical entities (Hull 1985a).

Marjorie Grene (1987) has objected to certain hierarchical treatments of behavior because in them the "actor" in "interactor" drops out altogether. But if conceptual change is construed as community-based, actors play several crucial roles in it. Not only are the brains of human beings important vehicles (in Campbell's sense) in conceptual replication series, but human beings are equally important inter-actors (vehicles in Dawkins's sense). They are among the chief vehicles for conceptual replicators. They are also the entities that juxtapose scientific hypotheses and natural phenomena in experiments and observations. Conceptual replication and interaction intersect in human agents.

Human beings also participate in the social relations that integrate individual people into communities. Science is inherently and necessarily a community affair. Certainly isolated hermits can learn about the world, but if science had been constituted in its early years by such hermits it never would have gotten off the ground (Hull 1985b). In order for science to be cumulative (to the extent that it is), transmission is required. Similarly, the sort of objectivity and rationality that gives science the peculiar features that it has are characteristics not of isolated individuals but of individuals cooperating and competing in peculiarly organized social groups (Hull 1978b, 1985b).

Biologists commonly note that entities at different levels persist for different lengths of time. One constraining factor on group selection is that the organisms that compose groups come into being and pass away so much more rapidly than the groups of which they are temporarily a part. With respect to calendar time, species speciate much more slowly than organisms reproduce themselves. Plotkin and Odling-Smee (1981) have extended this same observation to conceptual change. At each level in the relevant hierarchy, selection operates on a different time base. In this connection, the career lengths of particular scientists place some constraints on the speed of conceptual change. If we actually had to wait for aging scientists to die off before radically new ideas could become prevalent, this constraint would be prohibitive; however, no strong correlation seems to exist between age and the alacrity with which scientists adopt new ideas (Hull et al. 1978). Scientists change their minds on numerous issues during the course of their careers, but one thing is surely true: whatever a scientist is going to accomplish, he or she must accomplish in the space of a very few decades. Just when scientists get really good at doing what they are doing, they die. Individual scientists exist for a long time relative to the speed of conceptual change, but not long enough to encompass certain sorts of conceptual change. This is but another reason why scientific communities are important. In a more global sense, it is the continuity of scientific communities through time that allows for continued scientific change.

I once entitled a paper on sociocultural evolution "The Naked Meme." I ended that paper with the following cryptic observation: "If conceptual systems and their elements are interpreted as historical entities, actual transmission is essential, either directly from agent to agent in conversations or more indirectly through such means as the printed page. On the view being advocated by evolutionary epistemologists, conceptual evolution in the absence of social evolution leaves memes as naked as the apes who are their chief authors" (Hull 1982, p. 322). The main purpose of the present chapter has been to unpack this allusion by sketching the key role that actors play in selection processes by emphasizing how important interaction is. Omitting interaction in characterizing biological evolution leaves out the causal relations that make replication differential. Including reference to such relations but terming the entities involved *vehicles* makes them sound much too passive. Perhaps replication alone is adequate to capture the "bookkeeping" aspect of biological and conceptual evolution; however, in the context of scientific change, omitting reference to interaction leaves out not only reference to testing but also reference to the entities keeping the books – scientists.

NOTES

Reprinted with permission from *The Role of Behavior in Evolution*, edited by H. C. Plotkin (Cambridge, MA: MIT Press, 1988), pp. 19–50.

I would like to thank John Odling-Smee, Henry Plotkin, Elliott Sober, and G. C. Williams for reading and commenting on an earlier draft of this paper. Research for this paper was supported in part by NSF Grant SES-8508505.

1. Although Brandon (1982) and Sober (1984) agree that selection is a causal process, they disagree about which general analysis of causation can best handle selection adequately.
2. As much as Sober (1984) and Rosenberg (1985) disagree on other points, they agree that fitness is supervenient on the properties of individual organisms.

2

Taking Vehicles Seriously

Periodic reconceptualizing of areas of science to introduce greater coherence is as necessary as it is difficult. Wilson and Sober (1994) propose to restructure our understanding of selection processes centering on the concept of vehicles, not replicators, and they supply a means by which one can decide which levels of organization are functioning as vehicles at any one time. Although I find myself in basic agreement with Wilson and Sober's position, I think that one aspect of their exposition is extremely misleading. They write as if they accept Dawkins's (1976) distinction between replicators and vehicles, when actually they transform it significantly.

According to Dawkins, replicators pass on their structure through successive replications. Obviously, Dawkins developed his notion of a replicator with genes in mind, but I disagree with Wilson and Sober (1994) when they say that "all genes are replicators by definition." Dawkins's definition certainly lends itself to genes, but nothing in it requires that genes and only genes fulfill his requirements. Dawkins does go on to argue that genes are the only replicators, but he has to append these arguments to his definition.

In the beginning, according to Dawkins, the original replicators might have been naked genes, but eventually genes constructed vehicles of various sorts to aid them in their task of replicating themselves. For Dawkins, the connotations of the term "vehicle" are appropriate. He sees all-powerful replicators riding around in the vehicles that they construct and direct, discarding them for new survival machines when the occasion demands. Dawkins is certainly a gene *replicationist*, but he is also a gene *selectionist* only if he thinks that replication is not just necessary but also sufficient for selection. In his later publications, Dawkins (1982a, 1982b) makes it clear that he does not. Vehicles also

46

play an important role in selection. They form a hierarchy of increasingly inclusive vehicles. As everyone now seems to agree, the levels-of-selection controversy does not concern replicators but vehicles, or what I have termed "interactors" (Brandon & Burian 1984; Dawkins 1982a, 1982b; Eldredge & Grene 1992; Hull 1980; Lloyd 1988; Williams 1985).

Wilson and Sober (1994) treat Dawkins's "vehicle" and my "interactor" as synonymous, but as Dawkins and I have defined these terms, they are not. As Dawkins sees it, the primary relation between replicators and vehicles is developmental. Replicators produce vehicles of increasing inclusiveness. The replicator-vehicle distinction mirrors the genotype-phenotype distinction. As a result, replicators cannot themselves function as vehicles: replicators both produce and control vehicles but cannot themselves be vehicles. Genes cannot ride around in themselves, nor are they lumbering robots. Thus, if "vehicle," as Wilson and Sober use the term, allows genes to be vehicles and Dawkins's usage does not, Wilson and Sober must have transformed the meaning of this term in ways that remain only implicit in their exposition.

Although I cannot deny that the gene-organism relation also influenced how I originally conceived of the replicator-interactor relation (Hull 1988b, pp. 256–7), from the first I defined "interactor" in terms of the role that interactors play in *selection, not development*. As I defined this term, an interactor is an "entity that interacts as a cohesive whole with its environment in such a way that this interaction *causes* replication to be differential" (Hull 1988b, p. 134; see also Hull 1980, p. 318, and 1988a, p. 408). Thus, no entity is in-and-of-itself an interactor. An entity *functions* as an interactor depending on the role it plays in a particular selection process. An interactor must *cause* replication to be differential, but the causal mechanisms involved can vary. According to my definition, genes themselves can function as interactors. They can interact with their biochemical and cellular environments in such a way that they themselves are replicated differentially. Although Dawkins (1982b, p. 46) now doubts that the effects of genes need to be bundled together in discrete vehicles, he still conceives of vehicles as "units of phenotypic power of replicators" (Dawkins 1982b, p. 51). Dawkins is willing to talk loosely about phenotypic traits being "for" something (e.g., bird wings for flying), but not DNA. DNA codes for phenotypic traits, but DNA itself is not "for" anything. Even in Dawkins's extended usage, genes cannot function as vehicles.

According to Wilson and Sober, the "essence of the vehicle concept is *shared fate.*" The method that Wilson and Sober use to decide the level at which interaction is taking place has much in common with the "additivity" method first suggested by Wimsatt (1980, 1981) and later developed extensively by Lloyd (1988). I wish that Wilson and Sober had gone on at greater length about any similarities and differences that their method has in contrast to that of Wimsatt and Lloyd.

In sum, Dawkins's terminology lends itself to the claim that genes are the units of selection. Wilson and Sober counter that vehicles are the units of selection. According to the terminology I prefer, *there are no units of selection* because selection is composed of two subprocesses – replication and interaction. Selection results from the interplay of these two subprocesses. Genes are certainly the primary (possibly sole) units of replication, whereas interaction can occur at a variety of levels from genes and cells through organisms to colonies, demes, and possibly entire species. The units-of-selection controversy concerns levels of interaction, not levels of replication. However, to use the ambiguous term "selection" to characterize this controversy is to ask for continued misunderstanding.

In addition, I think that my concept of an interactor makes for a better restructuring of selection than does "vehicle"; but the term itself is also preferable, because "vehicle" has all sorts of misleading connotations (Brandon & Burian 1984; Eldredge & Grene 1992; Williams 1985, 1992): it implies hapless robots being controlled by all-powerful genes. Although this characterization may fit the entities that function in certain selection processes, it does not accurately characterize the entities that function in all selection processes. For example, demes may well function as interactors. They are in no sense "lumbering robots." Although colorful metaphors make for interesting reading, they have their costs as well. "Vehicles of selection" trips off the tongue all too misleadingly, unlike "interactors of selection."

NOTE

Reprinted with permission from *Behavioral and Brain Sciences* 17 (1994): 627–8. Reprinted with the permission of Cambridge University Press.

3

A General Account of Selection

Biology, Immunology, and Behavior

David L. Hull, Rodney E. Langman, and Sigrid S. Glenn

1 INTRODUCTION

What was so radical about Darwin's theory of evolution? In the following quotation, B. F. Skinner (1974, pp. 40–1) dismisses the usual
answers given to this question and suggests a nonstandard answer of
his own:

> Darwin's theory of natural selection came very late in the history of
> thought. Was it delayed because it opposed revealed truth, because it was
> an entirely new subject in the history of science, because it was charac-
> teristic only of living things, or because it dealt with purpose and final
> causes without postulating an act of creation? I think not. Darwin dis-
> covered the role of selection, a kind of causality very different from the
> push-pull mechanisms of science up to that time.

Although we think that the late appearance of selection on the intellectual scene no doubt had numerous causes, we agree with Skinner
that part of the answer is surely the counterintuitive kind of causality
exhibited in selection processes. Push-pull causation does seem
"natural" to us. So does functional organization. But the action of selection processes does not. This fact about how people in the West think
is reflected in natural languages. Finding terms to describe selection
processes that do not have all sorts of inappropriate connotations is
not easy.

Numerous biologists and philosophers of biology have presented
analyses of gene-based selection in biological evolution (e.g.,
Lewontin 1970; Dawkins 1976; Hull 1980; Sober 1984; Vrba & Gould
1986; Lloyd 1988; Sober & Wilson 1998), but relatively few workers
have tried to present a general account of selection to see which

49

processes in addition to gene-based biological evolution are genuine selection processes and which are not. (The chief exception is Darden & Cain 1989.) Are selection processes sufficiently different from other sorts of causal processes to warrant a separate analysis? The sort of selection that goes on in biological evolution is surely an instance of selection, but how about other putative examples of selection, for example the reaction of the immune system to antigens, operant learning, the development of the central nervous system, and even conceptual change itself (Cziko 1995)?

In this paper we provide a general account of selection. The chief danger of such "general analyses" is that they can be either too broad or too narrow. If the account is too broad, then everything becomes a selection process, including crystal formation and balls rolling down inclined planes. We have no objection to anyone attempting to present general accounts of more global phenomena, such as the persistence of patterns, but we limit ourselves just to selection processes. The other danger is to make the analysis too narrow so that each putative type of selection becomes unique. For example, the genotype-phenotype distinction plays a central role in gene-based selection in biology. Is this role common to all selection processes or unique to selection at the biological level? Being able to distinguish self from nonself is crucial in the immune system. Is the self-nonself distinction also important in other sorts of selection? In operant learning selection occurs only with respect to sequences of environmental interaction rather than with respect to numerous concurrent alternatives. Is this difference sufficient to disqualify it as a case of selection?

Such questions cannot be answered a priori. We have to try various analyses and see how they turn out. Our goal is to see if selection processes can be construed usefully as a special sort of causal process. The success of such an analysis will be determined by the use that those scientists working on various sorts of selection can make of it. If they find that our analysis helps them to understand the sort of selection they are studying more clearly, then it has succeeded; if not, then it has failed (for a defense of the method of abstraction, see Darden & Cain 1989). Even though causation is absolutely central to our understanding of selection, we do not attempt to present a general analysis of causation in this paper. In the past, some of the disputes that have arisen with respect to selection actually turn on different views of causation (e.g., Sober 1984, 1992; Brandon 1982, 1990; Brandon et al. 1994; van der Steen 1996; Glymour 1999). Ideally, we should include an analysis of "causation" alongside "selection." However, every analysis must stop somewhere. Not all of the substantive terms used in an analysis can themselves be analyzed. We do not present an analysis of causation in this paper because the literature is too vast and the alternatives too various. For better or for worse, in this paper we depend on the reader's largely tacit understanding of this extremely basic notion. The most that we can do in the space of a single paper is to point out when different notions of causation have caused problems, as in the instance cited above (for a recent discussion of causation, see Salmon 1998).

The three authors of this paper come from three very different backgrounds. David Hull (1980, 1987) emphasizes his work on gene-based selection in biological evolution, treating selection as an alternation between replication and environmental interaction. Sigrid Glenn contributes her work on operant behavior as a selection process (Glenn 1991; Glenn & Field 1994; Glenn & Madden 1995). Rod Langman (Langman 1989; Langman & Cohn 1996) adds his extensive theoretical analysis of the immune system. In this paper we strive to pool our conceptual resources to produce a general account of selection adequate for the three sorts of selection under investigation – gene-based selection in biological evolution, the reaction of the immune system to antigens, and operant behavior.

We do not offer an analysis of three other possible examples of selection – the development of the central nervous system, social learning, and conceptual change. We do not include an extensive discussion of neuronal development because the empirical facts remain too controversial (Edelman 1987; Quartz & Sejnowski 1997). Once neurophysiologists have worked out the basic structure of neuronal development, we then will be in a position to evaluate this process to see if it can legitimately count as a selection process. If it fits our analysis, well and good. If not, then either neuronal development is not a selection process or else our analysis is deficient. The second example of a putative selection process that we do not discuss in this paper is social learning, even though social learning is one of the most commonly cited examples of a selection process. Instead we limit ourselves to individual learning as a selection process. The strategy that we have adopted is to deal with the simplest cases first. Once we understand the most unproblematic instances of selection, we can then turn to the more difficult cases. The same justification applies to conceptual change, including conceptual change in science. Is the process that has allowed the three of us to understand selection in biological evolution, immunological reactions, and operant learning itself a selection process? Is conceptual change itself a selection process? Although we find the construal of conceptual change as a selection process fascinating, we do not discuss it in this paper (see Laland, Odling-Smee & Feldman 2000).

From the start, we have to register one warning: none of us claims to present the standard interpretation of the processes with which we are dealing, mainly because no such "standard" interpretation exists in any of the three cases that we investigate. For example, numerous objections have been raised to neo-Darwinian versions of evolutionary theory, especially the heavy emphasis placed on genes and the cavalier attitude frequently exhibited toward the environment. With a few noteworthy exceptions (e.g., Brandon 1990), the environment is treated as an unarticulated background against which selection operates. With respect to the immune system, considerable disagreement exists concerning the mechanism that allows the immune system to react selectively against nonself but not self components (e.g., Silverstein & Rose 1997). Numerous versions of learning theory can be found in psychology. Even if one limits oneself just to operant learning, disagreements exist. Can the stimulus that functions with respect to behavior be inside as well as outside the organism?

As much as scientists strive to reduce the amount of disagreement in science, they never come close to succeeding, and if science itself is a selection process, they cannot. In this paper we could not examine all versions of all of the theories that we treat. We had to select one from each of the domains. The issue is whether this version and others like it can be properly construed as selection processes, not whether we accept the reader's preferred version. This caveat applies with special force to theories of operant behavior. Some psychologists reject such theories out of hand. Others have strong preferences for one version over all others. In this paper we cannot answer the objections that have been raised to any of the three broad ranges of theories we discuss. Instead, for the purposes of this paper, we accept their overall adequacy and proceed from there to decide whether or not they exemplify a particular sort of process – selection. The point of this paper is not the choice of the one and only correct version of any of the theories that we treat. It is to discover if theories of this type can be construed as exemplifying selection processes.

Yet another problem that we confronted in writing this collaborative paper is that the three of us used very different terms to describe what we took to be the same sort of process. From the outset, we had to reduce differences that were mainly terminological, a task that turned out to be much more difficult than we had anticipated and not fully completed even now. One danger was allowing the process of biological evolution to play too large of a role in our undertaking. Because selection processes were first worked out in gene-based biological evolution, the temptation is to take it as standard and compare other candidates to it, but such a strategy would be biased. Historical precedence does not guarantee conceptual priority. In this paper we need to investigate each candidate in its own right, rather than taking gene-based selection in biological evolution as the standard by which all other putative examples of selection processes are to be evaluated. Even the use of the phrase "selection in gene-based biological evolution" is misleading. Both the functioning of the immune system and operant behavior are to some extent "gene based" and "biological." However, they also include processes that are not "gene based" in this narrow sense. But for want of a better name, we retain the phrase "gene-based selection in biology."

2   A BRIEF CHARACTERIZATION OF SELECTION

Several authors have attempted to characterize selection in as brief a fashion as possible. For example, Campbell (1974) describes selection as a function of blind variation and selective retention, while Plotkin (1994, p. 84) characterizes it as a matter of generation, testing, and regeneration. The trouble with these characterizations is that they are too brief. If one wants to understand selection, a sentence or two, no matter how succinct, will not do. Understanding space and time requires more than looking up these terms in a dictionary or in a physics text. Instead one must learn the relevant physics. Similarly, anyone who wants a deep understanding of selection has to study this phenomenon. Just inspection of a brief characterization of the process will not do. This much being said, we define selection as *repeated cycles of replication, variation, and environmental interaction so structured that environmental interaction causes replication to be differential.* The net effect is the evolution of the lineages produced by this process. Each word in this definition needs careful explication. The message is not to be found in the preceding brief characterization of selection but in the ensuing discussion.

2.1 Variation

Variation is sometimes considered part of the selection process (Darden & Cain 1989), sometimes as a precondition for selection processes (Hull 1980). Either way, variation is absolutely essential for the operation of selection processes. If there is no variation, then there are no alternatives to select among. However, the characterization of the variation that functions in selection processes has been one of the most contentious topics in the literature – and the most frustrating. It seems that no adjective exists in the English language that accurately reflects the sort of variation that occurs in selection processes. Is this variation blind, chance, random, nonprescient, nondirected, nonteleological, unforesighted, what?

First and foremost, the variations that function in selection processes of all sorts are caused – *totally* caused. No one writing in this literature feels inclined to introduce miracles in their descriptions of variation. The task is to describe the sorts of causes that produce this variation. When advocates of selection say that the variations that are operative in selection are "blind," they cannot possibly be using this term in a *literal* sense, as if some variations can see and others cannot. They must mean it in some metaphorical sense. When they term variations "chance" or "random," they cannot be using these terms as they are defined in mathematics. The requirements specified in these definitions are so rigorous that few, if any, natural phenomena can meet them.

Evolutionary biologists are well aware of the various factors that cause mutations. They are also aware that these mutations frequently depart from anything that might be termed "pure randomness." In fact, in many cases the very biologists who insist that the variations that function in selection processes are "random" are the ones who discovered these departures from randomness in the first place. For example, mutations that produce melanic forms crop up in certain groups of organisms with a greater frequency than the laws of chance would allow. On certain chromosomes hot spots exist that exhibit extremely high rates of mutation. For example, whole segments of immunoglobulin genes have bursts of mutation 10^6-fold greater than average (see Dawkins 1996, pp. 80–2; and Pennisi 1998 for additional examples). The other adjectives used to modify "variation" arise in the context of selection in conceptual change, but no one thinks that people, including scientists, are "prescient." People may try to anticipate the future, we can even predict the future in some cases, but no one is literally prescient.

Confusion in these matters stems in large part from the legacy of the early days of evolutionary biology, in particular the controversy between the Darwinians and Lamarckians. Critics of Darwinian versions of evolutionary theory tend to term any departures from the simplest forms of inheritance "Lamarckian." To be sure, numerous forms of nonstandard inheritance have been discovered over the years (Crow 1999). The issue is whether or not any of these fascinating forms of inheritance are in any significant sense "Lamarckian." The distinction between Darwinian and Lamarckian inheritance depends on the distinction between genotype and phenotype. According to the inheritance of acquired characteristics, the environment modifies the phenotype of an organism so that it is better adapted to the environmental factors that produced this phenotypic change in the first place – better adapted than those organisms that were not modified in this way. This phenotypic change is then transmitted somehow to the genetic material so that it is passed on in reproduction. Thus, according to this view, species can rapidly adapt to environmental change. In Darwinian evolution inherited variations are "random" with respect to (i.e., independent of) the effects that they produce, while in Lamarckian evolution they are not.

Both aspects of the preceding discussion need emphasizing. First, in Lamarckian evolution, the phenotypic change that results must make the organism better able to cope with the environmental factor that produced the phenotypic change in the first place. They must be adaptations. Exposing the skin to increased sunlight causes it to darken so that the organism is better able to withstand increased sunlight. Second, in Lamarckian evolution the phenotypic change must be transmitted to the *hereditary material* so that it can be passed on *genetically*. A mother dog giving fleas to her puppies is not an instance of the inheritance of acquired characteristics because it is not an instance of *inheritance* in the sense required by Lamarckian inheritance. Biologists do not have a corner on the term "inheritance." Other workers can and do use it in a variety of other senses. Our discussion, however, concerns Lamarckian inheritance as a biological phenomenon (for a sampling of the recent literature on Lamarckian forms of inheritance, see Lenski

55

& Mittler 1993; Jablonka & Lamb 1995; Rosenberg, Harris & Torkelson 1995; MacPhee & Ambrose 1996; Peck & Eyre-Walker 1998; Benson 1997; Andersson, Slechta & Roth 1998).

In sum, statements about the sorts of variation that function in selection processes need not include any reference to their being "blind," "random," or what have you. All of the terms that have been used to modify "variation" are extremely misleading. Hence, we see no reason to put any adjective before "variation" in our definition of selection. Our analysis concerns only those instances in which variations occur, without regard to their eventual contributions to fitness in biological evolution or some corresponding circumlocution with respect to the immune system and operant behavior. In this paper we deal with natural selection as it functions in Darwinian evolution today. Darwin himself included Lamarckian forms of inheritance in his theory, but Darwinians today do not. Darwinian evolution is currently limited to Darwinian (or Weismannian) inheritance. If Lamarckian forms of inheritance turn out to exist, we have no doubt that these mechanisms will be promptly incorporated into the Darwinian theory the way that neutral mutations were.

2.2   Replication

Replication is the second important notion in our brief characterization of selection, and it poses as broad a spectrum of problems as does variation. Replication contains two elements – iteration (or repetition or recursion, depending on one's terminological preferences) and information. Early on, Dawkins (1976) published a highly influential general account of selection that emphasized the role of "replicators." They are the entities whose structure contains the "information" that is passed on differentially in selection. The structure of replicators counts as information in the sense that it codes for the character of the individuals (or "vehicles") that the replicators produce. The only variations in the structure of replicators that matter are those that modify the relevant vehicles. These vehicles then interact with one or more local environmental conditions. Some of these variants survive to replicate and the process begins again. That is why Plotkin (1994, p. 84) in his analysis of selection emphasizes generation and *re*generation. However, sequential replication is not enough. Variants must be linked to proliferation so that at any one time, numerous alternatives are available for selection. At the very least, the frequency of replicators must change sequentially through time.

The only feature of the analysis of selection-type theories provided by Darden and Cain (1989, p. 110) with which we disagree is the demotion of iteration to an ancillary feature of selection. For them, selection is essentially a one-shot deal that can be, but need not be, repeated. They replace iteration with such evaluative notions as benefitting and suffering: "Several types of effects result from the differential interactions. In the short range, individuals benefit and suffer." Although they realize that such terms as "benefit" and "suffer" sound anthropomorphic and value-laden, they have to introduce them because they do not treat iteration as central to selection. A single cycle of replication and environmental interaction would fulfill the requirements of their analysis, just so long as it hurt or helped the relevant individuals.

In our analysis we are able to avoid the use of such problematic notions as "benefit" because of the central role of iteration. If some characteristic is increasing in frequency, then it is very likely (though not necessarily) doing some good. It is better adapted to its environment than other variants. According to our account, Darden and Cain's single-cycle analysis of selection is (at most) a limiting case of our account (see section 4.3 for further discussion). One reason why we prefer no mention of benefit and harm in our general account of selection is that their elimination from explanations of biological adaptations was one of Darwin's major achievements. We are not inclined to reintroduce such notions at this late date if we can avoid it. Iteration has problems of its own (e.g., how to keep "survival of the fittest" from degenerating into a tautology), but these problems can be handled with only a modicum of care and effort (see Lipton & Thompson 1988).

Replication is inherently a copying process. Successive variations must in some sense be "retained" and then "passed on." In many earlier definitions of "selection," all that is required is heritability, not genealogical inheritance. As Thompson (1994, p. 638) observes with respect to gene-based selection in biological evolution, natural selection "does not require genes or even direct descendants; all it requires is that the presence of a configuration of elements in one generation makes more likely the presence of the same configuration in the next generation." We agree with Thompson as far as genes are concerned but draw the line at descent. In biological evolution, replication is accomplished by molecules of DNA splitting and the missing nucleotides being filled in so that the information contained in the resulting molecules is retained. This is one way for replication to occur, but it is only one way. If splitting and reassembly is considered to be essential to all selection processes, then only gene-based selection in biological evolution and the functioning of the immune system count as selection processes. We think that this restriction is too narrow. A variety of mechanisms exist that can have the same effect as splitting and reassembly.

We have taken the opposite tack with respect to descent. Mechanisms other than modification through descent could serve the function that descent does. However, thus far, descent is the only mechanism that has evolved to produce the correlations necessary for selection. A more general analysis than ours might be couched in terms of retention of pattern or configuration from one generation to the next. However, in the absence of replication, the notion of "generation" becomes extremely problematic. In our analysis, we emphasize the mechanisms that produce evolutionary change, not just correlations. The preceding discussion is just one instance of the problems that arise in conceptual analysis. Is our analysis too broad or too narrow? Others might well make decisions different from ours, decisions that might have considerable merit.

In our analysis, the first component of replication is iteration (or repetition or recursion). The second is information. As Williams (1992, p. 11) points out, structure is necessary for selection, but structure alone is not good enough. Some of this structure must count as information. With respect to gene-based selection in biological evolution, "A gene is not a DNA molecule; it is the transcribable information coded in the molecule." DNA exhibits numerous structural elements. For example, it forms a double helix, and the bonds that connect the two bases that make up each of the rungs of the DNA ladder are easier to sever than those that connect successive nucleotides. With respect to gene-based selection in biological evolution, the preceding features of the DNA molecules count as structure but not as information. Of course, DNA itself had to evolve via selection. DNA molecules are adapted to replicate. The features of DNA molecules that allow them to replicate were selected for in the origin of life (Küppers 1990). But these features of DNA molecules do not "code for" anything.

In gene-based selection in biological evolution, much of the relevant information is composed of the linear sequence of bases in molecules of DNA. Unfortunately, in spite of the massive amount of work done by a variety of scholars on explicating the notion of information, none of the suggestions made thus far is adequate to distinguish information as it functions in selection processes from other sorts of structure. For example, physicists treat any structure as "information." The information contained in a double helix is not different in kind from that exhibited in the linear sequence of bases. As helpful as the work of Dretske (1981) and Küppers (1990) may be in other respects, it cannot be used to distinguish the special sort of structure exhibited by sequences of base pairs in molecules of DNA from structure as such. Nor is it adequate to make this crucial distinction with respect to the immune system and learning. The one bright spot on the horizon is that several biologists (such as John Maynard Smith) and philosophers of biology (such as Peter Godfrey-Smith) are currently working on the problem. Progress may be forthcoming. If we are to have an adequate conception of selection, progress in our understanding of information *must* be forthcoming. In the case of causation, the problem is that too many different analyses of causation exist, some adequate for certain causal situations, others adequate for others. In the case of information, the problem is that too few analyses of information exist, and none of them is adequate for understanding selection processes. In writing this paper, we were presented with two choices: register this major deficiency in our understanding of selection and move on or present from scratch an analysis of information that is up to the task. We decided on the first alternative. We hope that others will eventually come to adopt the second alternative (for a critical evaluation of the recent literature on information theory, see Sarkar 1996 and Harms 1998).

2.3   Environmental Interaction

Dawkins (1976) placed considerable importance on the notion of replication. It is the primary explanatory concept in his analysis of selection. Many critics think that Dawkins places too much emphasis on replication, as if it were sufficient for selection. They also raise the issue of the problematic character of information, as we have. Dawkins also introduced a second notion, that of a "vehicle." According to Dawkins, replicators replicate themselves (homocatalysis). In addition, they produce vehicles (heterocatalysis). Replicators do more than just cause or produce vehicles; they "code" for them. For Dawkins the relation between replicators and vehicles is that of development. A third major criticism of Dawkins's view of biological evolution turns on the relation that he sets out between replicators and vehicles. Replicators not only code for their vehicles but also ride around in and steer them. Vehicles are nothing but survival machines, lumbering robots controlled by the replicators that produced them.

To begin with, Dawkins's vehicles of selection have to be distinguished from Campbell's (1979) physical vehicles. For Campbell "vehicle" refers to the material basis or carrier of information; for example, molecules of DNA that incorporate information in the order of base-pairs, the paper on which books are printed, the plastic that was once used for phonograph records, and the chips in electronic computers. Clearly, Dawkins means something else by "vehicle." Most narrowly, he means the organisms produced by genomes. Needless to say, this narrow notion immediately raises the nature-nurture issue. In what sense does a genome code for an organism? A genome all by itself never produced anything (Marx 1995). Genomes plus numerous other factors produce organisms. However, according to the standard framework, both genes and environmental conditions *cause* traits, but only genes *code* for them. Of course, the metaphor of genes "coding for" traits remains as problematic as ever. We are well aware that sketches of several alternatives to the traditional gene-based view of biological evolution exist. Our concern in this paper is to provide an account of selection adequate for the traditional view, not to answer every objection raised to the traditional view.

If the general analysis presented in this paper is to be applied to specific instances, the terms used in this analysis must be "operationalized." Environmental interaction must cause replication to be differential. For example, drift is differential perpetuation without environmental interaction. In this connection, "selection for" is often distinguished from "selection of." A gene contributes to the development of a trait that interacts with the organism's environment so that this gene replicates more profusely than the genes of conspecifics that lack this gene and trait. This gene is being "selected for" this ability. A second gene adjacent to the first gene may piggyback on it. Because this second gene does not interact in the relevant sense with its environment, it is not part of the cause of this increase in frequency. As we construe selection, development is only *one* of the causal relations that can exist between what Dawkins terms "replicators" and "vehicles." For a truly general account of selection, a much broader relation is necessary. The relation must be causal, but it need not be developmental.

Numerous other processes are also operative. For example, molecules of DNA interact with their environments to replicate themselves, but this process does not involve anything like ontogenetic development in the production of vehicles.

If the distinction between replication and environmental interaction does anything, it goes a long way in resolving the levels-of-selection controversy. When Dawkins says that genes are the units of selection, he means replication. Genes are the primary units of replication and "hence" selection. When others such as Mayr say that organisms are the primary focus of selection, they mean environmental interaction. In gene-based biological evolution, organisms are the primary units of environmental interaction and "hence" selection. To be sure, both replication and environmental interaction are necessary for selection, but we do not think that either is sufficient by itself. Both are needed for selection to occur. As Lloyd (1988) has pointed out, the levels of selection controversy concerns environmental interaction, not replication. Entities from molecules of DNA, cells, and organisms to colonies, demes, and possibly entire species interact with ever more inclusive environments in ways that bias replication. Selection involves two processes and not one. There are units of *replication* and units of *environmental interaction*, but there are units of *selection* only in a highly derived sense, in the same derived sense that IQ is a measure of intelligence (Hull 1980; Heschel 1994).

3   SELECTION IN BIOLOGICAL EVOLUTION

The highly general characterization of selection set out in the preceding pages applies in a straightforward way to selection in gene-based biological evolution. In each case the sort of selection that population biologists study can be seen to be a special case of the more general analysis of selection provided in this paper. (For a recent criticism of analyzing selection in terms of replication, see Griesemer 1999.)

3.1   Mutation and Recombination

In gene-based biological evolution, the sources of variation are point mutations and recombination. Point mutations result in a single nucleotide being changed. Recombination results from the reorganization of the linear structure of DNA. As it happens, recombination produces most of the variation that is actually operative in biological evolution. The linear sequence of nucleotides in DNA provides the information necessary for the production of proteins. Any rearrangement of these orderly nucleotide sequences stands a chance of changing the genetic information encoded in its DNA and possibly the phenotype of the organism as well. The causes of variation in the genetic material are important. The effects that genes have on the phenotype of an organism are equally important. In selection, not only must genetic variations result in phenotypic variations, but also these differences must affect the individual with respect to survival and/or reproduction.

At one time biologists believed that the vast majority of mutations result in a decrease in proliferation, while only a small percentage increase proliferation or do not affect it at all (but see Peck & Eyre-Walker 1998). Mutations can fail to affect proliferation in two ways: either they have no phenotypic effects or else the phenotypic effects make no difference to survival and/or reproduction. Once biologists had more direct access to the genetic material, they discovered all sorts of unexpected things about it. They found that most of the genetic material has no apparent function. Perhaps it did in the past, perhaps its current functions have yet to be discovered, but right now most of the genetic material does not seem to do much of anything. In part as a result of the former finding, it turns out that most mutations are selectively neutral (i.e., as a result of environmental interaction, they neither increase nor decrease in frequency), while some are selected against (they decrease in frequency because of environmental interaction) and only a small percentage are actually selected for (they increase in frequency because of environmental interaction). As important as Kimura's (1983) work has proven to be, his claim that changes in our beliefs about the relative frequencies of these three types of mutation requires a "new" theory of evolution has not been widely accepted (see Brookfield 1995).

3.2   Replication

What are the primary replicators in biological evolution? Genes, larger chunks of the genetic material, and sometimes even entire chromosomes can function in replication. Replication at higher levels of organization may also occur, but the more inclusive the entity, the harder it is for the requirements of replication to be met. The important point is that once the notion of replication has been distinguished clearly from environmental interaction and selection, this question (and it is an empirical question) can be answered more definitively. What are the entities that interact with the environment in ways that result in differential replication? Everything from genes, cells, and organisms to hives, demes, and possibly entire species. Environmental interaction wanders up and down the organizational hierarchy, while replication is largely limited to the genetic material. In some circles the view that genes are the primary replicators and that environmental interaction occurs at a variety of levels is considered radical – possibly true but still in need of extensive elaboration and corroboration. In other circles, it is considered to be the received view that needs to be replaced by a more sophisticated theory.

Needless to say, wide agreement does not exist about the character of this more sophisticated view. As is usually the case in such disputes, one side parodies the other. For example, certain critics of the received view treat replication as a nonsense notion, as if replication is supposed to occur in the absence of any and all environmental contributions, but even the most rabid gene replicationist knows all of this. Quite obviously, replication requires all sorts of environmental inputs, including energy and the relevant enzymes (Marx 1995). Traditional versions of neo-Darwinian theory have enough faults without inventing irrelevant parodies. Perhaps evolutionary biologists have not spent enough time on attempting to integrate development into evolutionary theory, but they are well aware of its existence and the need for such an integration (e.g., Davidson, Peterson & Cameron 1995). Perhaps an adequate theory of evolution will require the sort of fundamental revisions that some critics of the received view suggest (e.g., Jablonka & Lamb 1995; and Griesemer 1998), but evolutionary biologists are likely to be swayed more by positive contributions than by continued criticism.

3.3   Environmental Interaction

In the traditional view of biological evolution, the primary means of recording (or retaining) and passing on variation is via genes. That is why we have been terming selection in biological evolution "gene-based." Then these genetic variants must interact either directly or indirectly with the environment so that in the last analysis replication is *differential.* Some replicates are more likely to be passed on than others. In addition to replicating, genes also code for phenotypes, and these phenotypes can be exhibited at various levels in the organizational hierarchy from genes, cells, and organisms to colonies, populations, and possibly entire species. Genes interact with their cellular environments, but they also interact with increasingly more complex environments via their surrogates. The "fit" between these phenotypes and their environments determines which genes get passed on and which not. In more general terms, the information contained in replicators gets passed on differentially because of how successfully they or their products interact with their respective environments (Brandon 1982).

What are the entities that function in environmental interaction? Can we get along just with the notion of phenotypic effect, regardless of these effects being bundled together into organisms? As strange as it might sound, genes themselves exhibit adaptations. The most obvious thing about DNA is that it is adapted to replicate. During periods of replication, genes interact with their immediate environments. They could not replicate without appropriate environmental contributions. Organisms exhibit phenotypic traits in the most obvious sense. Some organisms in a species have split telsa; others do not. At the other extreme, even species exhibit phenotypic traits. For example, the peripheries of the ranges of some species are highly convoluted. If speciation usually occurs at the peripheries of these ranges, then such convolutions, if they are heritable, might count as adaptations for increased rates of speciation. Some authors complain that requiring adaptations for selection, including species selection, is too restrictive. A more general notion is required, the sort of general characterization that we have provided (Lloyd 1988; Wilson & Sober 1994; Sober & Wilson 1998; and Gould & Lloyd forthcoming).

Much of the discussion of selection in the recent literature has concerned replication, but environmental interaction is at least as important in selection as is replication. The strongest feature of Darden and Cain's (1989) analysis of selection is the emphasis that they place on environmental interaction. As they put it, "individuals must be in an environment with critical factors that provide a context for the ensuing interaction" (Darden & Cain 1989, p. 110). The debate that Dawkins's *The Selfish Gene* (1976) prompted was generated in large measure by an ambiguous use of the term "selection" in the literature. One side of the dispute conflates "replication" with "selection," while the other side conflates "interaction" with "selection." Dawkins argues at great length that in biological evolution the relevant replicators are genes and only genes. Replication is certainly *necessary* for selection as it occurs in biological contexts, but it is not *sufficient*. Replication and variation in the absence of environmental interaction results in drift, and as important as drift may be in the evolutionary process, it is not a consequence of selection (Donoghue 1990). Selection requires an interaction of some sort between the environment and the replicating entity.

Dawkins's opponents have countered that organisms are the primary focus of selection. They, not genes, are the units of selection. Just as Dawkins, early on in the controversy, too often elided from replication to selection, his critics tended to equate selection with environmental interaction. As in the case of replication, environmental interaction is necessary but not sufficient for selection. Without replication, iteration is impossible, and in the absence of iteration, selection could not be cumulative. Selection is the result of differential replication *caused* by environmental interaction. Once again, selection is two processes, not one. It is the alternation of replication and interaction with the occasional introduction of variation.

3.4   The Environment

Of all the terms in the preceding characterization of selection, "replication" has received the greatest attention. However, the most difficult notion is that of the environment. Not until Antonovics, Ellstrand, and Brandon (1988) did it receive the analytic attention that it deserves (see also Brandon 1990). These authors distinguish between three different sorts of environment – the external, ecological, and selective environments. The external environment is the "sum total of the factors, both biotic and physical, *external* to the organism that influence its survival and reproduction" (Brandon 1990, p. 47). The ecological environment of an organism is composed of "those features of the external environment that affect the organism's contributions to population growth" (Brandon 1990, p. 49). Finally, the selective environment is an area (or population) that is "homogenous with respect to the relative fitness of a set of competing types" (Brandon 1990, p. 69).

One problem with respect to selective environments is whether or not to include other organisms, including conspecifics, as part of the selective environment. Such decisions have effects, for example, on how one handles cases of density-dependent population regulation (Brandon 1990, p. 65). Wilson and Sober (1994, p. 641) see this issue as clearly distinguishing their views from those of Dawkins. No sooner did Dawkins introduce the notion of a "vehicle" in his account of selection than he began to undermine it. "I coined the 'vehicle' not to praise it but to bury it" (Dawkins 1994, p. 617). According to Dawkins (1994, p. 617), "Natural selection favors replicators that prosper in their environment. The environment of a replicator includes the outside world, but it also includes most importantly, other replicators, other genes in the same organism and in different organisms, and their phenotypic products." Wilson and Sober (1994, p. 641) respond that Dawkins's goal of reconceptualizing vehicles of selection as part of the external environment (in Brandon's sense) reveals a deep contradiction in the gene-centered view of selection. Clearly the notion of environmental interaction deserves at least as much attention as replication.

4   SOMATIC SELECTION IN THE IMMUNE SYSTEM

More than a million different antibodies are needed to provide sufficient protection against the huge number of pathogens a host may encounter during its lifetime. Antibodies are protective because they act as markers that signal the recruitment of powerful biodestructive cells and enzymes which then destroy the pathogen and stop it from overgrowing the host. Managing to make sure that none of the millions of antibodies target any part of the host is obviously essential. If there were only a few antibody specificities, and a correspondingly small number of genes encoding these antibodies, then any rare cases of "self"-targeting might reasonably result in the destruction of that rare organism; this is an example of germline selection. When the rate of evolution of the pathogen (often hours) is much faster than the rate of evolution of the host (often months to decades), then the host genome cannot carry the millions of different genes needed to track the millions of different mutations in the pathogens. Moreover, among the millions of different antibodies, some will inevitably target a self component of the host and have the potential to destroy the host. What makes the immune system special is that it is able to select on the specificity of each antibody and eliminate the deleterious anti-self before it can actually kill the host. Because each different antibody is expressed in a different cell with a correspondingly different set of genes that encode that antibody, the immune system is able to select on the cell in order to eliminate these anti-self antibodies instead of having to eliminate the whole organism. This form of cellular selection on genetic variants is an example of somatic selection. In immunology it is common to refer to the germline as the genetic material that is selected upon when individuals are replicated and to distinguish this from the soma where the genetic material of individual cells can be varied and selected upon as cells are replicated. While the factual basis for phenomena discussed here can be found in any modern textbook of immunology, the conceptual analysis should not be taken as representing the standard view of the immune system.

4.1   Variation: The Origins of Antibody Diversity

The genetic basis of antibody diversity is partly due to the presence of several different, normally inherited genes, and partly due to mutations that occur in these genes when they are expressed in the soma as antibody-producing B cells. Extensive genetic and sequencing studies can be summarized along the following lines. The antibody molecule is made up of two different polypeptides, the L (light) and H (heavy) chains, which are encoded at two different genetic loci. The particular specificity of an antibody is determined by roughly equal contributions from the L and H chains. The part of each chain that is primarily concerned with antibody specificity is called the V (variable) region and the remainder the C (constant) region. Each region is encoded as a separate gene segment, and there are about 100 V-L and 100 V-H gene segments but only one C-L and one C-H segment. A series of gene fusions permanently changes the chromosomes in B cells and results in the joining of any one of the 100 V segments with the single C segment to produce a single V-C gene that encodes the complete L or H polypeptide. The 100 different L chains and 100 different H chains form random pairs and 10,000 corresponding different specificities. The gene fusions are arranged in such a way that joining errors are maximized. Consequently, few B cells are actually able to produce two L or two H chains. In other words, the B cell is made functionally haploid so that each B cell expresses only one kind of LH pair and, therefore, one specificity. Of course the level of waste is relatively high, as 70 to 90 percent of B cells that attempt to produce antibodies fail and are eliminated.

Throughout the life of an organism, the B cell population is undergoing constant renewal, and this renewal requires the mechanism for eliminating potentially self-reactive B cells to operate continuously throughout life. Controversy surrounds the details of this mechanism of self-nonself discrimination, but the exact nature of this mechanism is unimportant here. The result in any case has to be that individual B cells can be somatically selected according to the particular antigens that react with their receptors. The result is a means of selecting against B cells that can react with self components and neutral selection on B cells with specificities that do not react with self components. However, when a B cell that has *not* reacted with self is subsequently confronted by the particular pathogen with which it can react, then the B cell is strongly selected for, and, so long as the antigen persists, the cells undergo many rounds of mutation and division while secreting huge amounts of their antibodies. The negative selection pressure imposed by self components is constant (self is constant). During the many rounds of cell division that occur when a B cell is under selection by nonself antigens, mutations are introduced in the V segments of the L and H genes that, by chance, affect specificity. These mutations are so important that a special mechanism operates over the V gene segment and is able to introduce single base changes at the rate of 10^{-3} per base pair per generation; in contrast, the normal rate of mutation of around 10^{-9} per base pair per generation operates on the C segments. Some mutations in V segments are neutral and do not affect specificity, others destroy function, and a few change specificity and improve the ability of the antibody to react with its antigen at much lower concentrations than were present when the B cell was initially selected.

A brief comment on some terms and concepts might be helpful. Antigens are the parts of the pathogen that react with antibodies. Usually the term "paratope" is used to describe the part of the antibody that binds the antigen, and the term "epitope" is used to describe the site on the antigen that reacts with the paratopic part of the antibody. A complex pathogen, such as a bacterium, can expose many epitopes and induce the production of many paratopes, including mutant forms of the initially selected paratopes. The actual B cells that are selected by a particular antigen will depend on the concentration of the antigen and the affinity of the B cell receptor for that antigen. As a result, some B cells will respond only at high antigen concentrations while others will respond only at low antigen concentrations.

4.2   The Replication-Variation-
Interaction Sequence

The course of events following infection by a pathogen begins with a small inoculum of dividing pathogens. Initially they are at too low a concentration to cause the selection of any B cells. Then, after some time, the numbers of pathogens increases to reach a concentration that can induce an antibody response. The responding B cells proceed to divide and secrete huge amounts of antibody. As the antibody diffuses into the body fluids, it binds to the pathogen and so marks it for destruction. Providing enough antibody is present to halt the growth of the pathogen, the immune system will then have protected the host. As the numbers of pathogen decrease, the concentration of antigen driving the division and mutation of B cells also decreases with the net result that only those B cells able to respond to the lowest concentrations of antigen will remain dividing. In this case somatic selection for variant antibody genes allows some B cells to divide more often than others. During this period, B cells will have behaved in a manner very similar to the pathogenic organisms. Each will have also undergone mutation, and those pathogens that could render their antigens unrecognizable by the immune system will have been at a powerful selective advantage, whereas those B cells that could track the antigenic changes will help protect the host.

In summary, the B cell component of the immune system illustrates two levels of somatic selection. First, the steady flow of new B cells that can react against self components are deleted. Then, when nonself antigen happens to enter the host, those B cells that can react with the pathogen are selected to undergo many rounds of cell division and mutation with repeated selection for those B cells that continue to react with an ever decreasing concentration of antigen. This process is termed affinity maturation. When viewed in the context of the presence of environmental selection pressures (either self or nonself antigens), the individual B cells of the organism undergo a process that is indistinguishable from what is normally thought of as classical gene-based biological evolution of organisms, even though these B cells are not able to behave in all the ways often expected of an organism.

In terms of the overall effectiveness of the immune response, the affinity maturation process is of marginal significance because it occurs after the pathogen has been eliminated and can therefore only act during subsequent reinfections. The two significant selection processes occur first at the level of sorting the stream of new B cells into specificities that are either self (to be eliminated) and nonself (to be kept) and second at the level of amplifying only those B cells with specificities that react with the pathogen that suddenly and unexpectedly appears. The strict notion of serial rounds of replication, variation, and interaction applies only to the small component of affinity maturation in the overall immune response. However, it would be difficult to argue that the immune system does not undergo somatic evolution as a parallel to classical gene-based evolution found in the pathogens.

4.3   Somatic Selection versus Germ-Line Selection

Another important aspect of somatic B cell evolution is whether its origin as a part of the developmental program of the host is sufficient to disqualify this process as an example of selection. Included in this question is the inability of the immune system to continue evolving when the host dies. When the host dies of starvation or from being eaten by a tiger, it does not mean that the host's immune system is defective, it just happens to stop evolving because of some unselectable cosmic catastrophe. It seems unnecessarily restrictive to say that selection has to continue for some arbitrary period of time. To be able to show that somatic selection in the immune system stops for some reason other than a failure of the immune system is sufficient to conclude that a process of selection has been at work.

Many similarities exist between somatic evolution in the immune system and the functioning of the nervous system. In contrast to the detailed knowledge of the molecular and genetic structures and functions of the immune system, much less is known about the nervous system, and as our analysis of operant learning will clearly illustrate, even in a well-defined behavioral domain, the relevant molecular and genetic factors are almost unknown. Nonetheless, Edelman's ideas on immunology and neurobiology are sufficiently interesting to warrant comment. As a leading figure in the early years of modern immunology, Edelman was a strong proponent of what has come to be termed the big-bang version of the generation of antibody diversity. In particular he postulated a somatic genetic recombination mechanism that could generate a huge number of variants without having to resort to point mutations, which he thought to be rare and to occur throughout the genome (Gally & Edelman 1972). This initial burst of genetic diversification dispersed the variants in different B cells, which were then subject to selection with respect to self and nonself reactivity. Further genetic diversification after infection and antigen selection was thought to be minimal because it might include the introduction of new specificities able to react with and destroy self – the host.

Under big-bang models all of the diversity of the immune system is generated early in ontogeny, driven largely by the need to eliminate anti-self at an early stage in order to leave the remaining anti-nonself repertoire large and readily induced. This conceptual framework of big-bang diversification in the immune system can be found in Edelman's later work on neuronal development (Edelman 1987). Two points need to be made regarding big-bang and the immune system. First, big-bang diversification necessarily includes all the waste in all of the possible lineages selectable by antigen. Second, Darwinism, if it exists, must surely be more than selection from an unimaginably huge pile of possibilities. As we argue here, Darwinism involves multiple cycles of selection; that is, interaction, mutation, and replication. Big-bang requires only one round of mutation and selection, followed by continuous selection. While big-bang is difficult to justify when the immune system is constantly being regenerated, this is less of a difficulty in the case of the brain, where little cell division occurs once it reaches adult size.

The purpose of raising the big-bang principle is to emphasize that it denies the kind of serial selection we propose here for the three systems under investigation. Empirical observations notwithstanding, an a priori case can be made that if all possible variants are generated during big-bang and if the fraction of all possible variants used during the lifetime of an individual is very small, then the waste generated by unused variants is prohibitively large. *By generating variants among only those cells (or neural connections or organisms) that are already responsive to the selection pressure, waste in the production of unresponsive variants is limited and is not spread among all possible cells in all possible lineages. However, when variation is restricted to those entities responding to a selection pressure, each intermediate variant in a lineage must be individually selectable. If not, then the lineage would become extinct while waiting for a second or third variant to occur.*

One underappreciated selection pressure is the relative levels of waste, especially when evaluating probable versus improbable lineages.

Before accepting models of neural behavior based on big-bang diversification followed by somatic selection, it would seem prudent to consider possible alternative models based on serial selection, because the level of waste in the latter is likely to be substantially less than in the former case.

The widely quoted work of Hinton and Nowland (1987) provides another illustration of somatic selection gone awry. Their assumption of 20 switches in a neural network, each individually inactive, but providing a strong selective advantage when correctly coupled, is close to impossible. It may be true that given this impossible starting condition a form of somatic selection might be envisaged that is capable of selecting the right combination of switches, and that eventually a germ-line selection for the switches' all being in the right configuration is favored because the right combination is always found quickly. However, the exercise is rendered moot because the initial assumption is, at best, implausible. There is simply no remote likelihood of 20 gene duplication and mutation steps occurring in the absence of selection of the intermediates (i.e., the intermediate switches from 1 to 19 are individually unselectable).

4.4 Population-Level Selection in the Immune System

Several mechanisms have evolved to produce the massive amount of variation necessary to make the immune system work. The genes that code for antibodies have developed a variety of mechanisms needed to rapidly diversify a relatively small number of germ-line genes in a large somatic population of B cells. Central to these mechanisms is the generation of a functionally haploid genome in the region encoding antibody specificity. The introduction of mutational variants in these haploid specificity regions creates a population of different B cells, which are then subject to further individual selection by antigen. The immune system also exhibits a very different kind of variation that is uniquely expressed at the level of populations of host organisms. This variation is confined to the two to four genes that determine what is termed the Major Histocompatibility Complex (MHC) – the locus primarily responsible for the extreme difficulty in transplanting tissues. The MHC genes exist as a large number of alleles (about 100) that are found at roughly equal frequency in the interbreeding population. Although there are roughly the same number of alleles of the hemoglobin genes, all but a few alleles are at such a low frequency that they can be accounted for by mutation alone. To explain the roughly equal frequency of so many alleles at the MHC locus requires postulating a selection process that operates at the level of the genes in individuals and at the level of gene expression in the population.

The exact nature of the selection pressure operating on the MHC genes is not well known, but one compelling, illustrative explanation depends on the role that these genes play in immune protection against viral infections. In order for the immune system to respond appropriately to events occurring inside a virally infected cell (and without cracking the cell open to peek inside), the immune system uses the MHC genes to provide means for transporting intracellular peptides to the surface of the cell. Once these peptides are displayed on the cell surface, a special type of antigen-specific cell (the T cell) is able to bind specifically to the peptide under the right conditions and decide what to do. It works as follows: each MHC allele picks up a slightly different peptide fragment and presents it to the immune system. Viruses may well be selected for if they have mutations that disable or block the peptide binding site on the MHC and so stop the immune system from detecting the presence of intracellular virus. To combat this occurrence, the host has at least two, sometimes four MHC genes, each with a different peptide-binding specificity. If all individuals in the population had the same alleles with the same peptide-binding specificity, then, as the virus moved from one individual to the next, it could keep on evolving to defeat the MHC system. However, if the population possesses a large number of different alleles, then when the virus moves from one host to the next, all the selection in the previous host is canceled because the new host has new peptide-binding rules determined by the new MHC alleles. Thus, each allele functions perfectly well in an individual, but selection on the virus extends over many individuals at the population level. One result of this process is the large number of alleles in the population. This situation can be contrasted with the large number of antibody specificities needed per individual. The polymorphism of the MHC locus provides a particularly clear example of selection on alleles of genes that must occur at the population level while still being executed at the level of the individual organism.

4.5   *Serial Somatic Selection: The*
Immune System Is One Example

In this brief overview of the immune system, we have extracted four examples of selection. (1) In the case of the 100 germline-encoded V segments at the L and H chain loci, these segments can produce 10,000 different LH pairs with different antibody-binding sites. The selection that maintains these segments as different V segments is 100 pathogens that would be a threat if it were not for the specificities of these 100 unique LH pairs; the 9,900 other combinations are unselected as particular specificities and represent a very small form of "big-bang." (2) Among the unselected LH combinations and point mutants of these segments (up to one million of them), some are able to target self components of the host and have the potential to kill the host instead of the pathogen; these specificities are selected against by killing the cell that makes that specificity of antibody before the antibody is secreted and can kill the host. (3) Some specificities are able to target antigens of the pathogens, and the B cells that make these specificities are induced to proliferate and mutate so as to produce new specificities that function better (at lower concentrations of antigen) than others. This example of somatic selection is also an example of serial selection of the type that forms lineages akin to those found in the serial selection processes of the evolution of organisms. (4) In another domain of immune system function, MHC molecules play a critical role in allowing the immune system to be informed of the presence of pathogens located inside the cells of the host.

Although each individual organism (host) has two to four different MHC genes, in the population of organisms there are 50–100 alleles, all at roughly the same frequency; and this implies that selection, which must occur in individuals and their genomes, is via a selection pressure that only affects individuals because they are in a particular population (i.e., the selection pressure is particular to the population an individual finds itself a member of). This process is strictly germ-line selection, not somatic selection, and is sometimes referred to as "group selection." If we take a hard position on selection processes and require repeated rounds of replication, variation, and interaction, then the immune system offers one example of selection in affinity maturation. However, taken together, the other two examples of somatic selection also seem to simulate all of the features we might expect of serial germ-line selection in classical gene-based organisms.

5   OPERANT SELECTION

In one sense, all the behavior of organisms is the result of natural selection; in another sense, none of the behavior of organisms can be attributed to natural selection. The first statement follows from the fact that natural selection accounts for the range of behavioral potentialities characteristic of the organisms in any particular lineage and also for the processes that account for behavioral content that is uniquely suited to circumstances arising during an organism's lifetime. The second follows from the fact that processes other than natural selection are always involved when behavioral content actually appears in the behavior stream of a living organism. Between these two extremes lies the vast domain where behavioral scientists toil. Although no serious student of science would likely subscribe categorically to either of the two extremes, behavioral scientists with differing interests focus their attention on different segments of the continuum and tend to characterize those with interests elsewhere on the continuum as occupying one or the other of the extremes. Full scientific understanding of behavioral phenomena will require understanding the full range of behavior from one end of the continuum to the other.

5.1   Operant Behavior

When the behavior in which scientists are interested changes in content, often dramatically, during an organism's lifetime, one might say that those scientists are interested in the behavior of behavior. While such a locution sounds odd, a cursory look at how "behavior" is used in science reveals that scientists discuss the behavior of volcanoes, proteins, hurricanes, immune systems, and so forth. When change in the phenomena of interest is the object of scientific study, the scientists are said to be studying the behavior of the phenomena. If the phenomena of interest under investigation are the activities of organisms and those phenomena are themselves exemplified by change, then behavior change or the behavior of behavior is the object of scientific study.

The behavior changes of interest here are changes in behavior that occur during a single lifetime. The topic under discussion is further narrowed to those changes in behavior that result from a selection process that is conceptually parallel to the natural selection of organismic characteristics across generations of organisms. Most of the scientists studying this type of behavior designate it as operant behavior, and they designate changes in operant behavior of a particular organism "operant learning." Traditionally, operant behavior has been defined as behavior that operates on the environment and changes over time (in form, organization, or relations to the antecedent environment) as a function of "its consequences." From the present perspective "its consequences" is a shorthand way of saying the "goodness of fit between the behavior and consequent changes in the environment." In short, the particular operant behaviors that emerge and change during the lifetime of individual organisms are the results of "a second kind of selection" – a process that itself is the historical result of the "first kind of selection" (natural selection) (Skinner 1953, 1981).

Many questions arise from a selectionist characterization of operant learning. How does operant learning fit into what we know about the evolution of species by natural selection? How does this "second kind of selection" differ from selection processes that result in the origin (and history) of the species? What are the "units of selection" in operant selection? In this paper we address these issues briefly. We readily acknowledge that a complete explanation of operant behavior will involve processes other than operant selection, just as organic evolution involves processes other than natural selection. We also acknowledge that not all behavior is operant behavior and, hence, that no discussion of operant behavior will answer, or even address, all questions and issues pertaining to the range of phenomena in the domain of behavior.

We have chosen to focus here on operant behavior for both conceptual and practical reasons. Operant processes are known to occur in several phyla, suggesting that their origin reaches deeply into the history of life on earth. Second, and paradoxically, operant selection seems particularly relevant to humans (Schwartz 1974, p. 196). Hominid anatomical features such as opposable thumbs, a highly developed cortex, and vocal apparatus may have co-evolved with increasing susceptibility to operant selection. On the practical side, thousands of experiments have yielded a large and complex literature from which to draw, some even conceptualizing results in selectionist terms (e.g., Staddon & Simmelhag 1971). Unfortunately, we can only draw upon an extremely limited part of that literature and will selectively attend to work that clarifies the theoretical perspective presented here. Finally, operant behavior is the area of interest of one of the authors. In the same way that "a zoologist may specialize on vertebrates without denying the existence of invertebrates" (Dawkins 1983, p. 405), we do not deny the existence or importance of behavior that is not operant.

Operant behavior, like biological evolution, is one of those simple topics that are widely, persistently, and sometimes perversely miscon-strued or misrepresented (Todd & Morris 1992), thus guaranteeing that at least some readers will find what follows to be at odds with con-ceptions colored by such misrepresentation. In an attempt to preclude excessive cognitive dissonance, we begin with a few general points that we think critical to understanding the theoretical perspective presented below.

5.2 Adaptation and Complexity

In the larger context of biological evolution, an organism's operant behavior has the biological function of interfacing between the organ-ism and its world. An analysis of operant selection requires allowing the organism that behaves to recede to the conceptual background and making the interface itself the object of investigation. This change of perspective amounts to a figure-ground reversal from that which is apparent in direct perception. The new "figure" is by its nature difficult to "see" due to the temporal character of its structure. But behavior has structure of its own. It is made up of parts and wholes, which are parts of more inclusive wholes, and those parts have functions, as do the wholes. An operant repertoire is made up of interrelated behav-ioral lineages, each having its origin at a different time in the history of the organism, and each having its own history. As in the case of the evolution of life on earth, understanding the process requires focusing on particular lineages. Each behavioral lineage evolves in relation to its local environment, and changes in one lineage can impact other lin-eages in the organism's repertoire. A particular operant repertoire gen-erally becomes, over time, increasingly complex in terms of the number of lineages it comprises, the complexity of its component interactors, and the historical and ecological relations among them.

The processes by which operant adaptation occurs are viewed here as analogous to the processes by which biological evolution occurs. Specifically, operant selection (in concert with other processes) adapts organismic activity over time to "fit" the environment in which it occurs. If the environment moves out from under the behavior slowly enough, the behavior may be able to adapt to the changing environment. If the environment changes too rapidly, the behavior may be extinguished. As in the evolution of species, operant behavior fits the present environment because of past selection and not because of any future state of affairs. Further, operant behavior that is well adapted to its environment may not contribute to the survival of the organism that is behaving. For example, behavior that is well adapted for producing drug-induced euphoria may result in premature death of the organism. Operant processes work the same way whether or not particular behaviors are conducive to survival of the behaving organism. So far as survival and reproduction are concerned, operant behavior is a very sharp two-edged sword.

Gene-based selection is studied in bacteria and fruit flies as exemplars of a process assumed to account for all species. Similarly, operant selection often is studied in lever presses and key pecks as exemplars of a process that has been shown to operate with respect to more complex behavioral units. Most readers of this paper will readily accept the proposition that a single set of processes accounts for the structural and functional complexity of primates as well as bacteria. Although we trust they can entertain the analogous possibility that a single set of processes can account for structure and function of behavior far more complex than lever presses and key pecks, they may draw their dividing line between operant behavior and "higher" behavior wherever they please.

In the sections below, we provide examples of the ways in which operant selection results in behavior change. The theoretical language used to describe the process is the language we suggest for a general analysis of selection rather than the language used by the original researchers. We readily admit that we are viewing operant behavior from a nontraditional perspective and regret that some parts of the analysis are somewhat speculative. Evolutionary biologists had to develop evolutionary theory for decades in the absence of an adequate theory of heredity. Even after development of Mendelian genetics was under way, considerable time elapsed before these two groups of scientists were able to see how the theories could be combined into a single coherent theory. Operant researchers and neurophysiologists are in a comparable position today. Neural mechanisms are not well understood by most operant researchers and the ways in which operant behavior changes as it undergoes environmental selection are not well understood by most neuroscientists. Although experimental evidence supports a selectionist interpretation of operant behavior and its neural underpinnings, theoretical revision is likely to be required. On the positive side, massive experimental evidence supports a selectionist interpretation of operant behavior, and attempts to relate the findings to one another are increasing in number.

5.3   Operant Interactors and the Behavioral Environment

The relation between responses and consequent stimulation (environment) is the area where most operant researchers have focused attention. In operant selection, the primary role of entities traditionally identified as responses is that of "interactor," the unit which "interacts as a cohesive whole with its environment in such a way that this interaction causes replication to be differential" (Hull 1989b, p. 96). Although the most obvious entities functioning as interactors in behavioral selection are responses, some interactors in operant selection cannot easily be conceptualized as responses. For example, a group of responses may function as a cohesive whole in operant selection. The members of the group may be homogeneous, such as a burst of lever presses that interacts as a cohesive whole with its environment; or an interactor may be a cohesive whole made up of many different and functionally related parts, as in baking a cake or driving to work. Although the interactors that experimentalists work with in operant laboratories are often lever presses and key pecks, applied behavior analysts have demonstrated in hundreds of studies that operant selection occurs at many levels of behavioral complexity. To assist the reader in relating ideas presented here to previously acquired concepts, we will use "responses" for interactors that can easily be conceptualized as responses and use the more technically correct term "interactors" when the events are less easily conceptualized as responses in the traditional sense.

In operant theory, activity designated as a "response" does not require a stimulus (Skinner 1953, p. 64). Beginning with the assumption that a particular response occurs because it is elicited by another particular event is neither necessary nor helpful. Most operant responses are functionally related to stimulating events, but those relations are exceedingly complex. For purposes of exposition, we will concentrate on the least complex of operant lineages – those that might be compared to prokaryote lineages in biological evolution (Glenn & Madden 1995). Traditionally, these operant lineages have been called response classes, but that terminology raises the same conceptual difficulties that arose from calling a species a class of organisms (Glenn, Ellis & Greenspoon 1992). Both "response lineages" and "response classes," however, imply that operant responses are parts of a population and the characteristics of a population of interactors are the focus of our interest.

When operant behavior is seen as the figure against organism as ground, the elements involved in selection processes are analogous to (not the same as) those involved in gene-based biological evolution. In operant selection, the interaction step involves a relation between responses (interactors) in an operant lineage and changes in stimulation (consequences) that follow those responses. In the simplest example of operant selection, some relations between behavior and consequent stimulation have the effect of increasing the frequency of responses in the lineage to which the response belonged. This effect is called reinforcement. Other relations result in a decrease in the frequency of responses in that lineage. Depending on the nature of the change in stimulation, this effect is called either extinction or punishment.

The selecting environment (consequent stimulus changes) is a subset of a larger domain of events in the physical world that have function with respect to interactors (responses) in a particular operant lineage. The full range of environmental events having function with respect to the behavior of a particular organism (including events having discriminative, conditional, or motivating functions) is that organism's *behavioral environment*. Any behavioral environment is a subset of a still larger domain that comprises "the environment" as often construed – the physical world (including that part of it deemed "social"). These different uses of the word "environment" often go unrecognized and are the source of much confusion in the behavioral sciences, as they have been in the biological sciences (see Brandon 1990, ch. 2).

The facts underlying the points in the previous paragraph are incontrovertible, but the conceptual language calls for further explication. The relation viewed here as the interaction step in operant selection must itself be related to the concepts of variation, replication, and retention in operant behavior as one exemplar of our general analysis of selection. In the following sections, those concepts will be discussed in the context of further discussion of the ways in which an organism's operant behavior changes over time.

5.4   Response Frequency in
Operant Lineages

One of the earliest and most productive tools of operant researchers was the cumulative record. Although a record depicts only a small amount of information about each response recorded, it captures a critical feature of evolutionary processes – the frequency with which responses in a lineage appear over time. The responses depicted in the record are those that satisfy the contingencies of selection designed by the experimenter. By changing the selection requirements, researchers bring about changes in the frequency, distribution in time, and selectable properties of responses in an operant lineage. Such changes were the initial subject of research on schedules of reinforcement (Ferster & Skinner 1957). Each schedule specifies a particular kind of selection contingency and consequently each results in its own characteristic response distribution.

Although the selection process works at the level of single organisms and results in historical changes in operant lineages in that organism's behavioral repertoire, a schedule of reinforcement produces its characteristic distribution in different operant lineages of particular organisms, across organisms of a single species, and across species on planet Earth. The striking similarities in distributions of responses on a particular schedule in different operant lineages may be viewed as behavioral heteroplasties. That is, selecting environments having particular features result in operant lineages having characteristic distributions. The distributions arise again and again when the selecting contingencies are repeated.

5.5   Selectable Properties of
Operant Responses

All operant responses have in common certain properties, just as all organisms have in common certain properties. Common properties of organisms include length, width, height, and body mass. Gilbert (1958) identified the "fundamental dimensions" of operant behavior but, from the present perspective, he did not distinguish unequivocally between properties of operant responses and properties of operant lineages. Responses are components of individual lineages in operant selection, just as organisms are components of individual species (and lineages) in natural selection. Some demonstrated selectable properties of

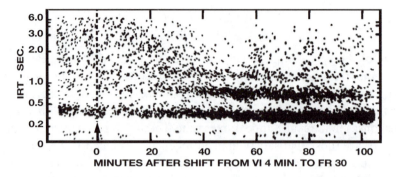

Figure 1. *Source*: From D. S. Blough (1963), "Interresponse time as a function of continuous variables: A new method and some data," *Journal of the Experimental Analysis of Behavior* 6:237–46. Reprinted by permission of Society for the Experimental Analysis of Behavior.

operant responses are duration, latency (interresponse time), force, form, direction, and relation to antecedent events. Because the level at which change is measured in selection processes is the population level, the properties of any one response are of little theoretical interest. Evolution occurs at the level of lineages and is measured in terms of response rate (frequency) or frequency of trait values in particular populations of operant responses.

Blough (1963) provided a graphic picture of change in an operant lineage as a function of change in selection contingency. In Figure 1, the response property of interresponse times (IRT) of successive responses in a pigeon's key-pecking operant are represented by the height of a dot on the ordinate of the graph. Time is represented on the abscissa. In the first 20 minutes, while a peck produced food only at the end of 4-minute intervals (VI 4 min. schedule of reinforcement), the distribution of IRT values in the population of responses was stable. The response population characterizing the lineage at that time shows a good deal of variation in IRT values (0.1 s–6.0 s), with a clustering of IRT values around 0.4 s. The vertical dashed line shows where the schedule changed to FR 30 (every 30th peck followed by food). During the next 100+ minutes, IRT values underwent a transition in which more and more IRT values clustered around 0.4 s, although variants continued to appear through the whole range of IRT values.

Such a change in population values in an operant lineage is conceptually equivalent to the often-cited change in coloration of successive generations of English moths undergoing anagenesis after industrialization. There are differences, however. First, the moth population (as in all sexually reproducing organisms) was distributed in space at any particular time and it extended in time across generations of moths. The distribution of IRT values in successive populations of pecks occurs only in the time dimension because organisms cannot press a lever more than once at a given time. Second, the trait of interest (IRT) in the operant lineage appears to have a broad range of values whereas the trait of interest (color) in the biological population appears to have a small number of discrete values. So at the resolution of human observation, responses appear to vary continuously at least in some of their dimensions.

The formal and temporal properties of responses in an operant lineage may vary widely within a population or they may vary within a narrow range of values. IRTs varied by a factor of 60 in Blough's data (Figure 1). This suggests that IRT may not have been the target of selection. That is, IRTs were not the property ("response trait") on which food was contingent but rather they changed along with other properties that were the target of selection. This distinction pertains to that between selection of multidimensional interactors and selection for their particular properties (traits) (see Sober 1984; Glenn & Madden 1995).

Although operant researchers have not traditionally presented data to demonstrate selection for particular properties, Catania (1973) depicted how response populations at successive times could be depicted to demonstrate the effects of selection for response properties of specified values. An adaptation of his graph for present purposes is shown in Figure 2. The changing frequency of force values in a lineage of lever presses is depicted as it could be observed in three populations of responses measured at successive time segments. Force values are fairly evenly distributed in the initial population of responses (A), representing a hypothetical population during a period of time before selection for specified force values. The curve labeled B represents a distribution of force values in a later population, after implementation of a selection contingency in which food pellets follow lever presses having force values between X and Y (and not otherwise). The dashed line represents the probability of consequent food for presses having force values between X and Y. The distribution of force values in the B population shows the effects of the contingency of selec-

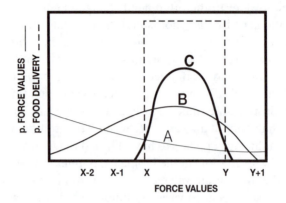

Figure 2. *Source*: Adapted from A. C. Catania (1973), "The concept of the operant in the analysis of behavior," *Behaviorism* 1:103–16. Reprinted by permission of Cambridge Center for Behavioral Studies.

tion on the operant lineage. Population C shows the effect of further selection for forces between X and Y. In the B and C populations, those responses falling outside the X–Y force range represent unreinforced responses (and would not appear in a cumulative record). Although successive populations in this graphical depiction of operant selection are increasingly composed of responses adapted to the selecting environment, variants that fail to meet the selection contingency continue to appear in the lineage, at varying frequencies at different times during the history of the lineage.

5.6   Variation in Operant Selection

Each interactor in operant selection has many properties and each property occurs at various values in responses forming a lineage. Rarely will two or more responses in a lineage be alike in all respects. Interactors with fewer components are likely to resemble each other more closely than will interactors having many parts, each of which can vary along many dimensions. When a behavioral interactor interfaces with the consequent environment, replication of all of its properties either increases or decreases in probability. However, only some of the interactor's properties may be required for an adequate "fit" with the selecting environment (e.g., the force values in the above example).

Those properties are differentially perpetuated in the population maintained by the current selection contingency.

Interactors in an operant lineage can be selected on the basis of the property of varying from their predecessor(s). Page and Neuringer (1985) performed a series of experiments in which a sequence of 8 pecks, distributed across 2 keys, was required to differ in their pattern from (a) the previous sequence (Lag 1), (b) the 5 previous sequences (Lag 5), and (c) other previous sequences up to 50 (Lag 50). In the present context, each sequence of 8 pecks is conceptualized as an interactor that either did or did not meet the requirements of the selecting environment. Selection was for interactors with a sequence of parts that differed from the sequence of parts of the interactor's immediate predecessor and for its last 5, 10, 15, 25, and then 50 predecessors. The 8-peck sequences showed variability in sequencing consistent with the selection contingencies, whether the requirements were gradually increased or whether a Lag 50 was implemented immediately (with other, naive, experimental subjects). Various control procedures demonstrated that interactor variation was itself being selected. In further experiments, the authors provided strong evidence that the variability observed was most likely randomly generated ("the pigeons behaved as a quasi-random generator," Page & Neuringer 1985, p. 447) rather than the result of some kind of memory function.

There is a difference in variation as a dimension of behavior and measures such as duration or force as dimensions of behavior. Variation is a direct measure of a population (like frequency or rate at which interactors are generated) whereas duration and force are measures of individual members of a population, which can be represented statistically as measures of a population undergoing selection (as depicted in Figure 2 above). Page and Neuringer (1985) concluded that variability in responses in an operant lineage is "a dimension of behavior much like other operant dimensions" (p. 450) in its susceptibility to selection. These and later experiments support that conclusion. And as in other kinds of selection, the susceptibility of variation to selection does not imply that selection is the *source* of the variation. Variants must occur before selection can operate. Page and Neuringer (1985) suggested that variation is an intrinsic property of operant behavior (i.e., has its origin in natural selection); operant selection can dial it up or down.

5.7 Origin of Operant Lineages

In one sense, as suggested earlier, all behavior has its origin in natural selection or, more proximally, in the inherited behavior of individual organisms. In some cases what is inherited has enough organization to be considered a behavioral lineage. For example, the pecking of pigeons is highly organized, in its formal properties as well as its relation to some properties of the environment, before operant selection begins to adapt features of an individual bird's pecking to local contingencies. Pigeons' pecking is a behavioral lineage that transcends the lifetime of individual pigeons. Its origin is in the history of the species. Operant modifications of the lineage during the lifetime of individual pigeons occur, but they are not encoded in the germline.

Some inherited behavior is not well organized with respect to its environment. Organisms of many species inherit a "supply of uncommitted behavior" (Skinner 1984). It is the kind of activity seen when an organism is in an environment that contains few elements with which it has means of interacting. Such behavior is prominent in the repertoires of human infants and can be seen on occasion in human adults (e.g., profoundly retarded adults or adults submerged in water or isolated for a long time in an empty room). The supply of uncommitted behavior is primordial in phenotypic behavioral development. Operant lineages emerge from the primordial behavior of a particular organism when selection is contingent on particular properties of the primordial activity and, as a result, those properties begin to appear more frequently in the behavior stream. If the selection contingencies gradually tighten, a response lineage gradually forms out of the more or less random (or, at least, poorly organized) activity.

Although the emergence of organized activity from undifferentiated movements can be seen to occur in real time, it has been difficult to study it experimentally because of a lack of equipment that allows recording of both the behavior meeting the changing contingencies and the rest of the behavior in the subject's ongoing behavior stream. Pear and Legris (1987) were able to develop a computer program that continuously tracked the position of a pigeon's head. They specified an arbitrary response (not seen previously during extensive observation of the pigeon in the experimental setting) as the experimental target. The pigeon's head was to make contact with a 3-cm-diameter "virtual sphere" at a particular location in the chamber. In addition to its precise spatial location, the form of response to be generated involved a dipping of the pigeon's head at that location. Beginning with a target virtual sphere that the pigeon's head would easily "contact," the experimenters gradually increased the frequency of movements making contact with the sphere and then gradually reduced the size of the sphere. As a result of these changing selection contingencies, the movements acquired the target form and occurred in that form at the target location at high frequency. The interactor lineage that emerged in each of the three pigeons' repertoires was maintained by stable reinforcement contingencies thereafter.

Operant lineages that exist in behavioral repertoires do not all arise from primordial behavior. Many operant lineages come into being by the splitting and merging of previously existing lineages. The complex relations currently studied in operant laboratories involve the merging and splitting of operant lineages (see Sidman 1994 for the history of one research program). Such behavioral complexity appears to require interactors that include stimulus parameters, interactors called "stimulus control operants" (Ray & Sidman 1970). The appearance of stimulus control operants in an operant repertoire has been likened to the appearance of eukaryotes in biological evolution (Glenn & Madden 1995). They allow for the grouping of responses into interactors having multiple parts and thus the evolution of behavioral complexity during the lifetime of one organism.

5.8 Replication and Retention in Operant Selection

Selection is a two-step process. "A process is a selection process because of the interplay between replication and interaction" (Hull 1981b, pp. 40–1). In operant selection, one step is the differential interaction between responses and consequent stimulation (environment) that "causes replication to be differential" (Hull 1989b, p. 96). The other step is the differential replication of response characteristics in successive generations. Whether operant selection is a process that parallels natural selection and belongs to the class of theories sometimes called "Darwinian" depends on the requirements one makes of the replication process. If the environment must have multiple and differing copies of a replicator concurrently available for selection to occur, operant behavior seems definitionally excluded. However, there appears to be no reason to assume that all replication processes involve concurrently existing events or objects. All that may be required is a process that retains features of interactors (event or object) across generations in a lineage, with a mechanism of variation to introduce novelty. As we said earlier, successive variations must in some sense be "retained" and then "passed on." This leads to questions regarding the site of retention of operant behavior and the mechanism by which "passing on" is accomplished.

So far as the material world is concerned, what is left after an operant interactor is gone is the central nervous system of the organism whose operant behavior is adapting to changing local contingencies. So the first step in operant selection occurs at the behavioral level (at the interface between organism and environment) and entails relations between interactors in a particular operant lineage and a selecting environment. And the second step occurs inside the organism at the neural level. Research on the biochemical mechanisms underlying learning and memory seeks to identify long-lasting changes that must occur in the strength of synapses as learning progresses and the learned behavior is maintained. "The range of possibilities for memory maintenance is large. None of the proposed models have been firmly excluded, and there seems to be no clear candidate" (Lisman & Fallon 1999). Full understanding of operant selection will require understanding of the relation between the two steps in the selection process. Because one step occurs at the neurochemical level and the other at the behavioral level, such understanding will necessarily entail synthesizing findings from research at these two levels. Unfortunately, researchers working on each of the two subprocesses, like geneticists and evolutionists before the modern synthesis, have little knowledge of one another's findings and often view with suspicion the conceptual framework of the other. There are exceptions. Donahoe and Palmer (1994) have begun to fashion a synthesis of biobehavioral processes in which they view neuroscientific findings in the context of a selectionist theory of learned behavior.

If the site of retention is the central nervous system of the learning organism, understanding of the mechanism(s) of retention will require investigation of changes in the properties (structural or functional) of neural activities as a function of differential interaction between responses and environment. In what follows, we draw on research that explicitly relates replication at the cellular level to operant processes. In a series of publications, Stein and his colleagues set out to assess Skinner's hypothesis that what constitutes a "response" at the behavioral level may not be that which is "strengthened" (i.e., replicated)

88

in operant selection. Rather, a response's "elements" or "atoms" (i.e., characteristics or traits) are the "units of behavior" susceptible to operant selection (Skinner 1953, p. 94). No such element can be correlated with a unit of replication smaller than a single neuron, so Stein et al. used in vitro preparations in which single neurons were subjected to analog contingencies of operant selection. For example, Stein, Xue, and Belluzzi (1994) made micropressure administrations of dopamine contingent on spontaneous bursting frequencies of single neurons of a hippocampal slice. They demonstrated that bursting frequencies increased when dopamine (a chemical associated with the reinforcing effect of drugs) was administered contingent on bursting, that the frequencies decreased when dopamine was not administered contingent on bursting and that bursting frequencies remained at or below baseline when they no longer administered dopamine independent of bursting (noncontingently).

In another study Stein and Belluzzi (1988) injected microadministrations of dopamine immediately *after* a postsynaptic neuron was activated by a presynaptic neuron, with a resulting increase in the presynaptic neuron's ability to activate the postsynaptic neuron. Other experiments (Self & Stein 1992) showed that it was not simply the stimulation of cellular activity that explained the effects of the burst-contingent dopamine. As suggested by Donahoe and Palmer (1994, p. 56), the effects of consequent dopamine on the ability of one neuron to activate another "demonstrates that dopamine can modulate the activity produced by glutamate, which is the major excitatory transmitter at synapses in the cerebral cortex, including those in the frontal lobes."

The work of Stein and his colleagues has several implications relevant to our analysis. First, the in vitro preparation demonstrated that the unit of replication is likely to be only a very small part of "complex neuronal circuitry associated with the reinforced response" (Stein, Xue & Belluzzi 1994, p. 156). A second, related implication is that the combination of cell firings can differ from response to response in a succession of responses of a lineage. Similarly, the combination of genes can differ in a succession of organisms of a lineage. Third, the in vitro preparation removes the operant selection process from any experiential requirements. As Stein, Xue, and Belluzzi (1994, p. 156) put the matter, "presumably, hippocampal slices do not experience 'highs.'" Fourth, the replication required for retention of interactor properties in an operant lineage may not require (but would not preclude) retention of a string of chemicals (as in DNA) across successive generations. Retention of operant properties in a lineage may instead be characterized in terms of the probability of a neuron activating other neurons of a pathway resulting in effector activity ("synaptic efficacy"). In vitro cellular analogs of operant conditioning further suggest that contingent reinforcement modifies several dynamical properties of a multifunctional network associated with motor behavior and that antecedent stimulation is not required for operant learning (Nargeot, Baxter & Byrne 1997, 1999). Presumably the mechanisms that account for retention of selected firing patterns will be found in the cellular chemistry of the modified network.

Differential interaction of responses and their consequent environments, then, has the effect of altering probabilities of the firing patterns of neurons. Differentially altered probabilities of events that "pass on information" (in this case, information coding for response properties) may be the hallmark of replication in selection processes. Unfortunately, we know little about the coding of information that is "passed on." Others will hopefully fill this knowledge gap. In sum, differential interaction of operant responses and their consequent environments causes differential replication of the properties of interactors in operant lineages. Researchers who consider themselves to be working in two different scientific domains (the "behavioral" domain and the "neuroscience" domain) have studied, respectively, the two steps of operant selection: interaction and replication. If researchers in both domains were to approach their work from a selectionist perspective and seek to synthesize their findings, a unified biobehavioral science of operant behavior would appear possible.

6   CONCLUSION

The goal of this paper is to present a general account of selection that is adequate for three putative examples of selection processes – gene-based selection in biological evolution, the reaction of the immune system to antigens, and individual operant behavior. After extensive reworking, sometimes generated by disagreements among the three authors of this paper, sometimes the result of two successive sets of referees' reports, we ended up defining selection in terms of repeated cycles of replication, variation, and environmental interaction. These three processes must be so structured that environmental interaction causes replication to be differential.

All three systems include variation. However, the respective amount of variation differs from system to system. For example, point mutations are introduced into genomes at very low rates, but these rates must be too high for selection because mechanisms exist that repair them. Enzymes roam up and down strands of DNA, seeking out "abnormalities" and repairing them. However, in gene-based selection in biology most of the variation is introduced by recombination, not point mutations. Rates of variation are extremely high in the immune system. More than one mechanism exists to make sure that the variation needed for selection is present in ample amounts. How variable operant behavior is depends on how finely we analyze behaviors. The natural units of variation are less obvious with respect to behavior than with respect to the other two systems (Enç 1995).

The most fundamental distinction made in this paper is between passing on information via replication and the biasing of this replication because of environmental interaction. As we have argued at some length, selection is not a single process but one composed of two processes – replication and environmental interaction. As a result, the issue of the levels at which selection occurs must be subdivided into two questions: at what levels does replication take place and at what levels does environmental interaction take place? These two questions elicit very different answers, depending on which of the three systems discussed in the paper is at issue. In gene-based selection in biological evolution, replication occurs primarily at the level of the genetic material, while environmental interaction takes place at a wide variety of levels, ranging from genes, cells, and organisms to kinship groups, demes, and possibly entire species. In the development of the immune system, gene-based selection in biological evolution plays the same role as in any other organismic system, but a second sort of selection also occurs. In somatic selection, those cells that specifically recognize a particular pathogen or foreign body respond and undergo extensive mutation and proliferation. Both replication and environmental interaction take place at the cellular level. In operant learning, the relation between an organism's responses and consequent stimulation causally affects the organism's central nervous system and subsequent behavior. The net effect is that some responses increase in frequency and others decrease.

Several problems arise in explicating the notion of replication. Even though the notion of "information" is fundamental to any account of replication, we do not provide such an analysis in this paper. We anticipate that in the future this need will be fulfilled. In replication the relevant information incorporated into the structure of replicators is "passed on" to successive generations of replicators. However, the mechanisms responsible for replication differ somewhat. Although the relevant replication in gene-based selection in biological evolution and the reaction of the immune system to antigens take place at the genetic level, the details of these processes differ. In addition, the distinction between self and nonself that is fundamental in the immune system does not play a corresponding role in the other two selection processes. In gene-based selection in biological evolution and the reaction of the immune system to antigens, genes replicate by splitting and filling in the appropriate nucleotides. Replication takes place in operant learning at the level of neurological processes, the nature of which still remains largely unknown.

Another difference that emerged with respect to these three instances of selection is between linear sequences of replication and their cotemporal proliferation. In operant learning, organisms react to sequences of events that result in cumulative changes – behaviors are reinforced or extinguished. However, in the other two forms of selection, extensive concurrent variations are presented to the environment. Although we think that a multiplicity of cotemporal replicators massively enhances the strength of those selection processes that incorporate such multiplicity, sequences of replicators that do not proliferate in this way also count as instances of selection, at the very least as a limiting case.

Environmental interaction is also necessary for selection. Some entities must interact with their environments so that the replication processes associated with these interactions become differential. Just differential replication alone is not enough for selection, that is, if such processes as drift are to be distinguished from selection. As in the case of information, we were confronted by the problem of distinguishing causal processes from other sorts of processes. In the case of information, none of the current analyses of information make the distinction necessary for selection processes. In the case of causation too many different analyses of causation exist, and none of them is totally superior to all others for all purposes. In all cases, however, selection consists of successive alternations of replication and environmental interaction.

The most common critical response to this paper will surely be that various researchers prefer different versions of the three theories than those that we have investigated; for example, replication occurs at levels higher than the genetic material, the mechanism that we have sketched for distinguishing between self and nonself is inadequate, or they simply do not like operant psychology no matter how it is formulated. But these objections are peripheral to the goal of this paper, which is to present a general account of selection that is adequate for the three sorts of theories that we have set out. Alternative versions of these three families of theories count against our analysis only if they cannot be characterized in terms of variation, replication, and environmental interaction.

If the preceding discussion has shown anything, it has been how counterintuitive selection processes actually are. The kind of causality involved in selection processes is, as Skinner (1974) noted, very different from our ordinary conceptions of causation. The two most striking features of selection processes are that they are both incredibly wasteful and yet able to produce genuine novelty and increased levels of organization. Given our ordinary notions, we might be led to ask how such wasteful systems can produce both novelty and increased organization. We suspect that selection processes are able to produce genuine novelty and organization only *because* they are so incredibly wasteful. The efficient production of novelty and order may not sound like an oxymoron, but we suspect that it is.

NOTE

Thanks are owed to Marion Blute, Todd Grantham, Allen Neuringer, Peter Richerson, Elliot Sober, David Sloan Wilson, Jack Wilson, and several referees. Langman was supported in part by NIH grant RR07716, and by the Programa PRAXIS XXI (16444), Ministério da Ciência e da Technologia, Portugal.

II

Selection in the Evolution of Science

4

A Mechanism and Its Metaphysics

An Evolutionary Account of the Social and Conceptual Development of Science

Scientists want credit for their contributions?
My dear, let us hope that it isn't true! But if it is true, let us hope
that it doesn't become widely known.

The idea of treating conceptual change as resulting from a selection process has seemed intriguing ever since Darwin established that biological species evolve primarily by means of natural selection. However, not until recently has anyone presented anything like a detailed application of the basic models of population genetics to conceptual change (Alexander 1979, Cavalli-Sforza and Feldman 1981, Lumsden and Wilson 1981, Schilcher and Tennant 1984, Boyd and Richerson 1985). Some of the models are literal applications. Various behaviors are treated as being genetically influenced phenotypic characters. Because my concern is conceptual change in science, literal applications are not liable to get me very far. We are a curious species. Our chief adaptation is playing the knowledge game. Our propensity to learn about the world in which we live is surely based in our genetic make-up. We might even be programmed to perceive the world in certain ways. For example, apparently people tend to divide up the visible spectrum along very similar lines, regardless of the contingencies of their culture (Berlin and Kay 1969, Bornstein 1973). Viewing the world essentialistically may also be part of our "hard wiring." If we are not born essentialists, at the very least we come to adopt this perspective quite rapidly.

We are also a social species. Our preference for living in herds is very likely to have some genetic basis. Since science itself is a social institution with its own norms and organization, our tendency to form groups also contributes to it. Even our general ability to use language

97

must have some genetic basis, and language is yet another prerequisite for science. In short, all of the prerequisites necessary for engaging in the process commonly termed "science" are to some extent programmed into us. However, relative to the generation time of our species, conceptual change has occurred much too rapidly for changes in gene frequencies to have played a significant role. Hence, conceptual change in science may be a selection process, but it cannot be gene-based. Changes in gene frequencies are liable to have very little to do with the specific content of particular scientific theories. The mode of transmission in science is not genetic but cultural, most crucially linguistic. The things whose changes in relative frequency constitute conceptual change in science as elsewhere are "memes," not genes (for the literature on evolutionary epistemology, see Campbell 1974, Bradie 1986, Campbell et al. 1987, Plotkin 1987).

In the past most authors who have treated cultural evolution in general and scientific change in particular as selection processes have taken gene-based natural selection as the exemplar and reasoned analogically to conceptual change. However, a more appropriate strategy is to present a general analysis of selection processes equally applicable to all sorts of selection processes (Toulmin 1972, Campbell 1974). After all, the reaction of the immune system to antigens is an instance of a selection process which differs just as radically from gene-based natural selection as does conceptual change in science. Any analysis of selection processes must apply to it as well as to natural selection. Such a general analysis must be sufficiently general so that it is not biased toward any particular sort of selection process but not so general that any and all natural processes turn out to count as instances of selection. Biological evolution, the reaction of the immune system to antigens, and cultural learning must count, but not lead balls rolling down inclined planes or the planets circling the sun.

Even if such a general analysis of social learning as a selection process is successful, it alone is not enough to account for conceptual change in science. The most puzzling feature of science is that it works so well in realizing its manifest goals, so much better than any other social institution. By and large scientists really do what they claim to do. All social institutions exhibit norms, but even when these social norms are translated from their usual hypocritical formulations to accord more closely with the norms that are actually operative, individual infractions are common. In science they are correspondingly rare. In 1985 at a meeting of the American Association for the Advancement of Science, William F. Raub, Deputy Director for Extramural Research and Teaching at the National Institute of Health, presented a paper in which he reported detecting only fifty cases of serious fraud out of 20,000 projects supported by the institute during a five-year period. Most of these cases involved such things as sloppy record keeping, the rest outright falsification of data. Observations in science are certainly, to some extent, theory laden, but the experiments which scientists run are not just *pro forma* rituals designed to support preconceived conclusions. By and large, scientists really do run the experiments that they claim to run, present their findings accurately enough, and are influenced by the results. They even acknowledge the contributions which other scientists make (for additional estimates of the amount of various sorts of "fraud" in science, see Zuckerman 1977; Culliton 1983, 1986, 1987; Norman 1984, 1987; Marshall 1986; Koshland 1987; and White 1987).

Although we are currently experiencing a period of science bashing, even the most hysterical critics have been unable to come up with rates of "fraud" in science that begin to come close to those repeatedly documented in other social institutions. Either scientists are not guilty of the substantial amounts of fraud common in other professions or else they are extremely good at covering it up. The romantic view of scientists as dispassionate, disinterested seekers after truth for its own sake needs debunking, but not because science is a plot against "The People," not simply because it is overly idealized, but because there is some danger, though not much, that scientists might actually be tempted to put this romantic view into practice. If scientists at large adhered to the professed mores of science, science might be possible, but I doubt it. As it is, scientists have too much sense to behave too strictly in accordance with the way that officially they are supposed to behave.

Numerous authors before Darwin suggested that species might evolve. A few even suggested mechanisms for this evolution. Darwin was so much more efficacious than his predecessors for a variety of reasons. He not only suggested a mechanism to account for the transformation of species but also worked it out in great detail. It is one thing to sketch the general outlines of natural selection, as both Wallace and Darwin did in their Linnean Society papers. It is quite another to explain how the extraordinary characteristics of sterile castes in eusocial organisms might have resulted from natural selection as well as dozens of other problem cases. Darwin was especially adept at transforming anomalies into confirming instances.

In addition, Darwin presented massive amounts of data to support his theory, much of it obtained from the work of others, and set out his views in a scientifically respectable way. The *Origin of Species* makes for slow reading because it contains so many examples, so much data, but Darwin did not stop with the *Origin*. He spent the next two decades publishing book after book in which he investigated one aspect of evolution after another, each laden with masses of data. Darwin's theory might be wrong, but no one could claim on Darwin's death that he did not take it seriously. More than this, he intertwined his professional career with those other scientists, scientists who stood to gain or lose depending on the fate of Darwin's revolutionary theory. Strangely enough, Darwin's presentation of natural selection as the primary mechanism for the transmutation of species legitimized the idea of organic evolution even though most of his contemporaries rejected it as the chief mechanism for evolutionary change (Ghiselin 1969, Ruse 1979).

The standards for scientific acceptability have hardly decreased since Darwin's day. Anyone who expects an evolutionary account of conceptual development in science to be taken seriously must attempt to meet these standards. In this short paper, however, all I can present is a sketch, along with the promissory note that in my book (Hull 1998a) I set out a more detailed account along with the necessary data.[1] Instead of explaining such things as the slavemaking habits of ants, I have to explain why scientists on occasion behave so altruistically, for instance, by giving credit to their closest competitors, and why on occasion they do not. Under what conditions are scientists likely to give credit to other scientists, and under what conditions are they likely to claim priority for themselves? As one might imagine, I do not think that "intellectual justice" is the entire story. Scientists do strive to behave the way that they "should," but other factors are operative as well, in fact more operative.

Much of what I have to say about the mechanisms that drive science has been said before. The two innovations which I introduce to explain the behavior of scientists are *conceptual inclusive fitness* and the *demic structure of science*. Just as organisms behave in ways which result in replicates of their own genes or duplicates of these genes in close kin being transmitted to later generations, scientists behave in ways calculated to get their views accepted as their view by other scientists, in particular those scientists working on problems most closely associated with their own. Scientists also tend to organize themselves into tightly knit though relatively ephemeral research groups to develop and disseminate a particular set of views. On the account that I am urging, conceptual change in science should be most rapid when scientists are subdivided into competing research groups. The factionalism that scientists themselves so often decry facilitates rather than frustrates progress in science (Blau 1978).

Science is inherently both a competitive and cooperative affair. If all that mattered was a scientist's individual conceptual inclusive fitness, the relationships between scientists would be complicated enough, but when alliances and allegiances are added, the story gets even more complicated. However, these complications cannot be ignored because science *is* essentially social. Individuals can learn about the world in which they live by confronting it directly, but if science is to be cumulative, social transmission is necessary. In addition, the sort of objectivity that gives science its peculiar character is a property of social groups, not isolated investigators.

CONCEPTUAL INCLUSIVE FITNESS

Science is a matter of both cooperation and competition. Both sorts of behavior are natural; both need explaining. How come scientists can cooperate as much as they do in such a competitive activity? The most important sort of cooperation that occurs in science is the use by one scientist of the results of the research of other scientists. The worst thing that one scientist can do to another is to ignore his or her work; the best thing is to incorporate that work into one's own with an explicit acknowledgement. Somewhere in between is use without explicit acknowledgement. Scientists want their work to be acknowledged as original, but just as importantly, they want it to be accepted. To get it accepted, scientists must gain support from other scientists. One way to gain support is to show that one's own work rests solidly on preceding research, but the price of this support is a decrease in apparent originality. One cannot gain support from a particular work unless one cites it, and this citation automatically both confers worth on the work cited and detracts from one's own originality. Scientists would like total credit and massive support, but they cannot have both.

Science is so organized that scientists are forced to trade off credit for support.

Several factors influence which way a scientist opts in particular circumstances. For example, scientists whose support is worth having are likely to be cited more frequently than those whose support is worth little or nothing. The preceding tendency is accentuated by the fact that there is little point in omitting reference to the contributions of a famous scientist because everyone who counts already knows who the author of this contribution is. When one fails to cite a well-known author, one gains neither credit nor support. The same observations do not hold for those lower in the scientific hierarchy, in particular one's graduate students. In the short term, their support is worth very little. However, graduate students are not entirely powerless because they are likely to be the chief conduits for one's work to later generations. Their success increases one's own conceptual inclusive fitness. Parent-offspring conflicts should be as common in science as elsewhere, and the resolutions of such conflicts should have the same general features.

Just as organisms in general behave in ways likely to increase their own genetic inclusive fitness, scientists tend to behave in ways calculated to increase their own conceptual inclusive fitness. In neither case are the entities involved necessarily aware of what they are doing. Flour beetles are unaware that such a thing as genetic inclusive fitness even exists, let alone capable of performing the required calculations. Scientists are not appreciably different in their quest for conceptual inclusive fitness. The functioning of science had better not depend crucially on widespread self-awareness among scientists about their own motivations or the effects of their actions because most scientists are no more reflective about the ongoing process of science than are other professionals about their professions. Self-delusion rules.

So, we are to believe, hospitals are run primarily for the good of patients, universities are run primarily for the good of students, and governments exist primarily to serve the people. Anyone who believes any of the preceding claims knows very little about hospitals, universities, or governments. Patients, students, and citizens do get some benefit from the relevant institutions, but when conflicts of interest arise, they are not always reconciled in the favor of those at the bottom of the hierarchy. In fact, they rarely are, self-serving rationalizations of the most embarrassing sort notwithstanding. However, I have found that scientists are a good deal less resistant to looking at their profession realistically than are members of other professions, primarily because even when the hypocrisy and romanticism are stripped away, science still retains its traditional characteristics. Science does have the characteristics traditionally attributed to it but not for the reasons usually given.

By definition, professions are occupations which are self-policing. One thing is clear about professions: by and large they do not police themselves very well, at least not according to the criteria that they profess. However, scientists police themselves as if science were made up of nothing but neighborhood yentas. In all social institutions both individual and group interests exist. They need not conflict, but all too often they do. We are all constantly being urged to behave in ways that are for the good of some group or other. Such appeals do have some effect, but too often not enough. University professors really should devote more time to undergraduate teaching. Those of us who are actively engaged in research periodically feel guilty about not spending more time on teaching and produce all sorts of justifications whose effect is to make us feel less guilty, but that is about all. Scientists adhere to the norms of science so well because, more often than not, it is of their own best self-interest to do so. By and large, what is good for the individual scientist is actually good for the group. The best thing that a scientist can do for science as a whole is to strive to increase his or her own conceptual inclusive fitness.

Such striving is kept within certain bounds by two factors: the need of scientists to use each other's work and the possibility of empirical testing. However, in order to evaluate the manner in which these factors function in science, two ways in which a scientist can "sin" against the mores of science must be distinguished. The first is by publishing faulty work, whether intentionally or unintentionally ("lying"). The second and most common is the failure to give credit where credit is due ("stealing"). Scientists "lie" when they publish views either which they know to be false or else for which they have failed to perform the work necessary to warrant their making these views public. In everyday life, guilt is a function of intent. Getting in a car and intentionally running someone down is much worse than intentionally getting drunk and unintentionally running someone down, even though the person may be equally dead in both situations. Officially at least, a parallel distinction exists in science. Publishing fabricated data is the worst sin that a scientist can commit. Publishing the results of sloppy research is considered to be bad but somehow not quite so bad, even though the effects on anyone using fabricated and sloppy research are indistinguishable.

One explanation for this distinction in science is that science is not an activity that lends itself to rote procedures. Scientists cannot publish the truth and nothing but the truth because, at the cutting edge of science, no one can tell for sure what this truth is. To be successful in their investigations, scientists must be adept at judicious finagling. Time and again in the course of their investigations, scientists must exercise judgment. Sometimes these decisions are patent. If one spills coffee over a culture of mouth protozoa and they all die, no one would object to a scientist omitting this data point. But other decisions are not so obvious. Should one correct for the age of the culture? Perhaps young cultures are more (or less) resistant to stannous fluoride than older cultures. Scientists try to account for or discount as many of these contingencies as they can, but time, money, energy, and insight limit this activity. The boundary between understandable error and inexcusably sloppy work is quite fuzzy. All scientists know that at times they have erred in this respect. Although those scientists who blatantly fabricate data are well aware of what they are doing, the boundary between error and fabrication can also become fuzzy. In certain circumstances, the only way to distinguish between the two would be extensive psychoanalysis. Why are Mendel's data so much better than they should be? Did he consciously fudge the results of his investigations or did he unconsciously classify borderline cases to enhance the relation between expected and observed values? All the efforts of the Mendel industry notwithstanding, we are unlikely ever to know.

Intent to one side, publishing work that other scientists use and find to be mistaken is punished severely in science, much more severely than "stealing," i.e., trying to pass off someone else's work as one's own. It is also, by all indications, much rarer. Why is lying so much rarer in science than stealing? Because it is punished so much more severely. Why is it punished so much more severely? Because stealing hurts only the person whose work has been appropriated, while lying hurts anyone who uses this work. Misassigned contributions are just as useful as work whose authorship is attributed correctly. However, the preceding comment concerns single instances. If the misassignment of credit in science were to become commonplace, the system would be seriously jeopardized.

The career of Sir Cyril Burt is a good case in point. When other scientists thought that all he had done was appropriate to himself the

104

work done by his assistants, no one was especially excited. After all, that is what assistants are for. But when it began to appear that he had fabricated not only these assistants but also their research, his fellow scientists became more than a little anxious because it brought into doubt all of the work that they had published which was based on his fabricated results. Fabricated results need not be any more mistaken than the results of sloppy research, but there is no reason to expect them to be correct either. As science has been conducted in the past couple of centuries, by and large scientists have been rewarded fairly consistently for doing what they are supposed to do and punished just as consistently when they transgress the actual mores of science.

As important as individual conceptual fitness is in science, conceptual "demes" also play an important role. Because few scientists have all the skills and knowledge necessary to solve the problems that they confront, they tend to band together to form research groups of varying degrees of cohesiveness. One function of these research groups is the sharing of conceptual resources (Giere 1988). These demes tend to be extremely ephemeral. They form and dissipate before anyone is liable to notice that they exist. However, every once in a while, one will seem to have made some sort of headway or breakthrough. A flurry of activity ensues, generally with little effect. Most activity in science, whether individual or group, has little discernible effect on science (for a critical review of this literature, see Fox 1983). However, on occasion, one of these groups is successful in the sense that others notice its achievements and either refute or adopt them. When the former occurs, interdemic selection is superimposed upon individual inclusive fitness. When the latter occurs, a set of views that originate in a small research group becomes widely disseminated, and interdemic selection is replaced by mass selection. As a result, only those research groups in science that are successful or significant failures are liable to be noticed.

Scientists do not simply read the literature to discover the truth. Rather they read it with an eye for work that bears on their own research. If a particular finding supports their own research, they are liable to incorporate it without testing. Testing is reserved for those findings which threaten one's own research. This tendency is accentuated in research groups. The scientists most likely not only to adopt one's views but also to be harmed most if they prove to be mistaken are one's allies. Some commentators on science urge scientists to test any and all work before they utilize it. They insist that a scientist should personally check everything that goes into any paper that bears his or her name (Broad and Wade 1982). I cannot conceive of worse advice. The whole point of scientists working together is to pool conceptual resources. Warm feelings to one side, cooperation in science is behaviorally indistinguishable from mutual exploitation. Perhaps one worker is mathematically quite adept but has very poor hands, while another can contribute very clean data sets even though he or she has to take the mathematics on faith. The best advice for a scientist who begins to doubt the reliability of the work of a colleague is to sever professional ties. One cannot waste one's time checking the results of one's coworkers.

Initially, criticism and evaluation come from within a research group. After publication it shifts to scientists outside the group, in particular to one's opponents. Individual scientists are to some extent objective. They know that, if they are lucky, their work will be held up to scrutiny. Hence, they best expose it to severe tests prior to publication. But each of us is also, to some extent, a prisoner of our own conceptual system. We take some things so much for granted that it never even occurs to us to question them. We also have our own career interests. On occasion scientists have refuted the very views for which they are famous, but not often. More often the really severe testing comes from one's opponents. It is also the case that one's opponents are liable to have different though equally unnoticed presuppositions. The self-correction so important in science does not depend on scientists being totally unbiased or having no career interests, but on other scientists having different perspectives, not to mention career interests. Scientists working outside your own research group are hurt if they adopt any of your mistaken views, but more importantly, they are also in a better position than you are to expose them to severe tests. *Their* career interests are not damaged if *your* views are refuted.

The preceding has concerned the effects of intra- versus intergroup lying. The effects of intra- versus intergroup stealing are just as dramatic. One's closest research associates are most vulnerable not only to the effects of the quality of one's research but also to having their findings appropriated. As science is now structured, scientists need not make their work available to other scientists until they publish and, hence, establish intellectual ownership. The major exception is the refereeing process. It is possible for a referee to read a manuscript, extract an idea, and get it published in time to gain priority. Because manuscripts submitted for publication are well down the road to completion, stealing of this sort is relatively difficult. Research proposals, to the contrary, are supposed to be largely prospective in nature. They are supposed to describe current research and likely avenues for future work. Hence, reviewers of research proposals have access to the plans of their closest competitors early enough so that they might well be able to gain credit for them before their authors can. As a result, scientists have devised various techniques to obtain research money without giving their competitors too much of an edge. But there is no way to hide one's work from others in one's own research group. Priority disputes between individuals working in relative isolation from each other are vitriolic enough. Those among scientists belonging to the same research group are even more devastating.

Lakatos (1971) has suggested that one way to decide between different views of science is by how many characteristics of science flow naturally from it and how many remain anomalies. The key unit in science as far as Lakatos is concerned is the research program. Research programs in turn are evaluated as to whether they are progressing, stagnating, or degenerating. Because these features of research programs are a function of recognized accomplishments, priority disputes between advocates of different research programs are a matter of "rational interest," while those between scientists promoting the same program result merely from "vanity and greed for fame." Although no one has collected systematic data on this topic, my impression is that priority disputes are as common among scientists working in the same as in different research programs. All of the former are, for Lakatos, anomalies. On my view of science, two processes are involved: individual inclusive fitness as well as intra- versus interdemic selection. Scientists should feel a strong allegiance to their allies for a variety of reasons, some of them nobler than others. Among the less noble is that one's own self-interest is tied up with the interests of one's co-workers. As long as the effects of individual and demic selection coincide, no problems need arise. However, when an individual perceives that his or her individual inclusive conceptual fitness is decreasing in part because of his or her participation in a particular scientific deme, internal friction is guaranteed to increase. On my view, priority disputes both within and between research programs are equally a matter of rational interest.

One might grant that the preceding is a reasonably accurate description of how science as we know it happens to function but nevertheless lodge two objections: first, that this structure is just an historical accident stemming from the general characteristics of the societies in which science happened first to emerge, and second, that science would fulfill its traditional goals even better if its structure were changed. Science did arise in the West in highly competitive, individualistic societies in which property rights were paramount and then diffused to other countries, some of which were, at the time, organized quite differently. If only science had been able to emerge on its own in these other countries, it might have exhibited very different properties than those that happen to characterize it now.

This hypothesis is certainly plausible. I have only two responses to make to it. First, science arose independently several times in the West. In several instances, most notably the French Academy, considerable effort was expended to organize science along more genuinely cooperative lines so that scientists could work in relative anonymity to promote the general good. In every case these efforts failed and were replaced by the system I have described (Hull 1985c). Appeals to work diligently for the general good have never proven to be powerful enough. The greatest strength of science as it is now organized is that it harnesses our "baser" motivations for more "lofty" goals. As the members of the French Academy officially acknowledged in their revised constitution in 1699, in the future instead of working in consort, each member of the Academy "shall endeavor to enrich the academy by his discoveries and improve himself at the same time."

If biology has anything to teach us about functional systems it is that there are always many ways to skin a cat. Functional equivalents are pervasive. Perhaps dozens of different ways exist for the organization of science which would promote our understanding of the natural world. So far only the barest sketches have been presented for these alternative forms of organization, and the few pilot studies that have been conducted have failed decisively. Perhaps the way that science is now organized is far from ideal. It may even be the case that in certain areas competition has become so fierce and the numbers of people involved so large that traditional mechanisms are breaking down. Biomedical research is the best example of science at its worst. Whether these rogue areas of science can be brought back into the fold by more careful attention to traditional mechanisms or whether an entirely new system must be introduced, I for one cannot even begin to guess (for one recent example, see Connor 1987).

SELECTION PROCESSES

Thus far, I have argued that science is a function of the interplay between cooperation and competition for credit among scientists. I have yet to say anything about its selective character. The literature on selection processes in biology has been plagued by a systematic ambiguity in the phrase "units of selection." Some insist that genes are the primary focus of selection because they are the entities that pass on their structure largely intact from generation to generation. Others insist that organisms are the primary units of selection because they are the entities that interact with the environment in such a way that genes are replicated differentially. Still others view the interplay between these two processes as "selection" and define "fitness" accordingly (Brandon and Burian 1984).

For the purposes of a general analysis of selection processes, terms such as "gene," "organism," and "species" are not good enough. Instead more general terms are needed. More general terms are needed if conceptual change is to be viewed as a selection process, but they are also needed in biological contexts as well. If the traditional organizational hierarchy of genes, cells, organs, organisms, colonies, demes, populations, and species is taken as basic, then in point of fact the focus of "selection" wanders from level to level. More than this, traditional entities do not function in the immune system the way that they do in biological evolution, and it is as much a selection process as is biological evolution. For example, selection as it functions in immune reactions takes place entirely in the space of a single generation, and the results of particular selection regimens are not passed on genetically. Selection in immune reactions is ontogenetic, not phylogenetic.

Because such traditional entities as genes, organisms, and species do not consistently fulfill the same roles in biological evolution, not to mention immune reactions and conceptual change, more general units are needed, units that are defined in terms that are sufficiently general to apply to all sorts of selection processes. My suggestions for these units and their definitions are as follows:

replicator – an entity that passes on its structure largely intact in successive replications.

interactor – an entity that interacts as a cohesive whole with its environment in such a way that this interaction *causes* replication to be differential.

With the aid of these two technical terms, selection can be character-ized succinctly as follows:

selection – a process in which the differential extinction and proliferation of interactors *cause* the differential perpetuation of the replicators.

Replicators and interactors are the entities that function *in* selection processes. Some general term is also needed for the entities that are produced *as a result of* replication at least and possibly interaction as well:

lineage – an entity that persists indefinitely through time either in the same or an altered state as a result of replication.

In order to function as a replicator, an entity must have structure and be able to pass on this structure in a sequence of replications. If all a gene did was to serve as a template for producing copy after copy of itself without these copies in turn producing additional copies, it could not function as a replicator. Although genes are well adapted to function as replicators, it does not follow from the preceding definition that genes are the only replicators. For example, organisms also exhibit structure. One problem is the sense in which organisms can be said to pass on their structure largely intact. For example, changes in the pellicle of a paramecium are passed on directly when the organism undergoes fission. From the human perspective, populations do not seem to exhibit much in the way of structure, but population biologists recognize something that they term "population structure." If popula-tions can pass on this structure during successive replications of these populations, then they too might function as replicators (Williams 1985).

Many cohesive wholes exist in nature, but only a few of them func-tion in selection processes. Hence, only a very few count as interactors. In order to function as an interactor, an entity must interact with its environment in such a way that some replication sequence or other is differential. Organisms are paradigm interactors. They are cohesive wholes, they interact with their environments as cohesive wholes, and the results of these interactions influence replication sequences in such a way that certain structures become more common, while others become rarer. However, many other entities also function as interac-tors. For example, genes not only code for phenotypic traits, they themselves have "phenotypes." DNA is a double helix which can unwind and replicate itself. In doing so it interacts with its cellular environment.

In the beginning, one and the same entities had to perform both functions necessary for selection. Because replication and interaction are fundamentally different processes, the properties which facilitate these processes tend also to be different. None too suprisingly, these distinct functions eventually were apportioned to different entities. When a single structure or entity must perform more than one function, it usually performs none of them very well. Too many compromises have to be made. Interaction occurs at all levels of the organizational hierarchy, from genes and cells, through organs and organisms, up to and possibly including populations and species.

As I have characterized it, selection is an interplay between two processes – replication and interaction. Both processes taken separately and the interplay between them are causal processes. As a result, drift does not count as a form of selection – as it should not. An entity counts as an interactor only if it is functioning as one in the process in question. Thus, if changes in replicator frequencies are not being caused by the interactions between the relevant interactors and their environments but are merely the effects of "chance," then the changes are not the result of selection. Drift is differential replication in the absence of interaction.

Many entities persist indefinitely through time. Of these, some change while some do not. However, the only entities that can count as lineages in the technical sense which I am proposing are formed by sequences of replicators. Hence, on my usage, "lineage" is inherently a genealogical concept. For example, the solar system has changed through time. Nevertheless, it does not count as a lineage because no replication was involved in such changes. The general notion is that of an historical entity – a space-time worm. Lineages are historical entities formed by replication. Differential perpetuation caused by interaction is not necessary for something to count as a lineage. In fact, differential perpetuation itself, regardless of its causes, is not even necessary for something to count as a lineage. However, when the interplay between replication and interaction causes lineages to change through time, the end result is evolution through selection.

Both genes and organisms form lineages. In most cases gene lineages are wholly contained within organism lineages. According to more gradualistic versions of evolutionary theory, species do not *form* lineages. Instead they themselves *are* lineages. But nothing about the evolutionary process requires that it be gradual. It might be the case that species are incapable of indefinite change and that speciation is always saltative. If so, then particular species themselves do not count as lineages. Instead, successions of species form lineages. It might be the case that not all organisms belong to species. For example, if significant gene exchange is necessary for species to exist, then they did not exist for the first half of life on Earth and are still absent among a significant proportion of organisms living today. Even though species are not a necessary consequence of replication, lineages are.

THE ROLE OF INDIVIDUALITY IN SELECTION

By now, one thing should be clear. Everything involved in selection processes and everything that results from selection are spatiotemporal particulars – individuals. Both replicators and interactors are unproblematic individuals. To perform the functions they do, they must have finite durations. They must come into existence and pass away. Replicators must exhibit structure, and interactors must interact with their environments as cohesive wholes. These are the traditional characteristics of individuals. That individuality is at the heart of the selection process can be seen by the frequency with which biologists who want to argue that species can be selected begin by arguing that they have the properties usually attributed to individuals, ordinary perceptions and conceptions notwithstanding (Eldredge and Gould 1972). Conversely, those who argue against species selection begin by arguing that species lack these very characteristics. In fact, those who argue that not even organisms can function as units of selection begin by casting doubt on their status as individuals, superficial appearances notwithstanding (Dawkins 1976).

Lineages are also individuals but of a special sort. In order to function as a replicator, an entity can undergo minimal change before ceasing to exist. In order to function as an interactor, an entity can undergo considerable but not indefinite change. Lineages are peculiar in that the organization which they exhibit is sufficiently loose so that they can change indefinitely through time but sufficiently tight that the effects of selection are not lost. Thus, any entity that can function either as a replicator or as an interactor cannot function as a lineage because they are too tightly organized. Conversely, any entity that can function as a lineage cannot function as a replicator or as an interactor because it lacks the requisite internal cohesiveness. All sorts of factors can enhance or destroy the cohesiveness of a lineage, but genealogical connectedness through time is essential.

All of the preceding may seem excessively fluid, but this fluidity is dictated by the nature of living entities. According to what we know of the functioning of the genetic material in both autocatalysis and heterocatalysis, genes are anything but beads on a string. Organisms come in a dismaying variety of forms, some well-integrated, others not. Even though colonial forms of organization depart significantly from our vertebrate perspective, they are widespread. Some organisms form colonies; others not. Some species are composed of demes; others more homogeneous, and so on. Any adequate theory of biological evolution must apply to all organisms, not just to well-integrated sexual organisms. As strange as the notions of tillers and tussocks, genets and ramets may sound to a zoologist, plants evolve too (Jackson, Buss, and Cook 1986).

What is more, entities do not stay put as they change through time. For example, in certain colonial wasps, each hive initially includes several queens. As time goes by all are killed off but one. As a result, the focus of selection expands from the individual organism to the entire hive. In a large, genetically heterogeneous species in which crossover is common, very small segments of the genetic material are likely to be the primary replicators, but when such a species goes through a populational bottleneck, entire genomes may come to function as single replicators. Similar observations hold for lineages. In any system that is evolving through selection, a point occurs at which the genealogical fabric is rent and networks become trees. At whatever level in the traditional hierarchy that this occurs, the resulting entities are lineages. For example, among strictly asexual organisms, no lineages exist that are more inclusive than organism lineages. Among sexual organisms, some form lineages no more inclusive than sequences of populations. In some species gene exchange may be sufficiently extensive and sustained to integrate entire species into lineages. Among some plants, gene exchange among traditional taxonomic species may be sufficient to integrate them into a single lineage (Mishler and Donoghue 1982).

All of the preceding may seem not only too fluid but also wrongheaded. Genes and organisms are unproblematic individuals, while species are just as unproblematic classes. After all, the terms "organism" and "individual" are interchangeable. First off, common usage may be the place where all investigations are forced to begin, but it cannot be where they end as well. At times, ordinary usage is misleading and must be modified. Secondly, all organisms may count as individuals according to ordinary usage, but not all individuals are organisms. "Individual" refers to a much broader spectrum of entities than organisms even in ordinary usage, including stars, continents, buildings, and nations. Furthermore, when one tears one's attention away from vertebrates and looks at all living creatures, many organisms turn out not to be paradigmatic individuals. Many biologists do assume uncritically that all organisms are individuals and vice versa, but they are wrong to do so.

Such issues to one side, some of my readers may take the distinction between spatiotemporally localized individuals and spatiotemporally unrestricted classes (or sets) to be of no consequence. Once again, terminology intrudes. I am not in the least interested in the terms that are used to mark this distinction. I take the distinction to be important, not the terms. The distinction is important because it has characterized science throughout its existence. Whether or not scientists have been mistaken to do so, one of the fundamental goals of the most important scientists in the history of science has been the discovery of natural regularities that apply to any entities whatsoever, just so long as they meet certain conditions. According to some, the claim that Moses wandered in the Sinai or that all the coins in my pocket are dimes differs in no important respects from Newton's law of universal gravitation. If so, then we have all been behaving in extremely inappropriate ways because we pay very little attention to those who come up with trivial claims of the first sort and heap fame and fortune on those few scientists who produce statements of the second sort. As difficult as it may be to present a totally satisfactory analysis of the distinctions between singular statements, accidentally true universal generalizations, and laws of nature, these distinctions are fundamental to our understanding of science.

Implicit in the preceding discussion is the distinction between two sorts of entities: those that are spatiotemporally restricted in the relevant sense and those that are not. Perhaps for philosophical purposes, Moses, all the coins in my pocket, *Dodo ineptus*, gold, and bodies with mass are all equally "sets." If so, then I am forced to distinguish between two sorts of sets: those that must be spatiotemporally restricted and localized to perform the roles they do in the natural processes in which they function, and those that must be spatiotemporally unrestricted to perform their quite different roles. To function as a replicator or an interactor, an entity must be spatiotemporally localized and cohesive, while lineages must be less tightly organized but just as spatiotemporally localized.

Once the preceding distinctions are made, decisions as to which entities belong in which category are largely a contingent matter. Do all genes and only genes function as replicators in the evolutionary process? The answer is clearly no. Do all organisms and only organisms function as interactors in the evolutionary process? Once again, the answer is clearly no. How about species? Given the groupings of organisms commonly considered species by systematists, do all and only species form lineages as a result of biological evolution? Although others might think otherwise, I think that the answer to this question is just as clearly no. Traditionally, particular species such as *Dodo ineptus* have been treated not simply as classes but as natural kinds akin to gold and triangle. However, if species are considered as chunks of the genealogical nexus, then they are as much spatiotemporally restricted and localized as are organisms and genes. Perhaps one might not want to term species "individuals," perhaps one might want to introduce a third category to the traditional distinction between individuals and classes for lineages (assuming species count as lineages), but species cannot be construed as spatiotemporally unrestricted regardless of whether they function in biological evolution as replicators or interactors, or only result from the action of selection processes operating at lower levels (Ghiselin 1974, Hull 1976, Mayr 1987). Common sense to one side, genes and organisms are not unproblematic examples of spatiotemporally localized individuals, and species are not unproblematic examples of spatiotemporally unrestricted classes.[2]

SCIENCE AS A SELECTION PROCESS

All of the preceding concerns biological evolution. If any regularities are to be found in evolutionary processes, general conceptions such as those I have set out are necessary. Perhaps my particular conceptions will turn out not to be good enough, but if both biological evolution and the reaction of the immune system to antigens are to count as selection, then traditional conceptions will not do. More general conceptions are necessary even in strictly biological contexts. My concepts have the added virtue that they are sufficiently general to apply to conceptual evolution as well, in particular to conceptual selection in science. None too surprisingly, the replicators in science are elements of the substantive content of science – beliefs about the goals of science, the proper ways to go about realizing these goals, problems and their possible solutions, modes of representation, accumulated data reports, and so on. Scientists in conversations, publications, and lectures broach all of these topics. These are the entities that get passed on in replication sequences in science. Included among the chief vehicles of transmission in conceptual replication are books, journals, computers, and of course human brains. As in biological evolution, each replication counts as a generation with respect to selection.

As they function in the production of proteins, genes are organized into reasonably discrete, hierarchically organized functional units. Because crossover does not respect the boundaries of these functional units, the chunks of the genetic material that are transmitted intact are highly variable. In biological evolution, these variable chunks of the genetic material are the primary replicators. According to some authorities, organisms and even possibly gene pools can also function as replicators in the evolutionary process. In any case, genes both have phenotypes of their own and code for more inclusive phenotypes which influence their perpetuation by means of their relative success in coping with their respective environments. Conceptual replicators interact with that portion of the natural world to which they ostensibly refer no more directly than do genes with their more inclusive environments. Instead they interact only indirectly by means of scientists. Scientists are the ones who notice problems, think up possible solutions, and attempt to test them. They are the primary agents in scientific change.

Conceptual replication is a matter of information being transmitted largely intact from physical vehicle to physical vehicle.[3] In addition to accuracy of transmission, these vehicles have two important characteristics – their duration and how active they are. Certain vehicles are much more ephemeral than others. The spoken word is extremely transitory; human beliefs incorporated into individual brains can last for longer periods of time, but people die. If a belief is to survive, it must be replicated. In books and journals, ideas find a much more durable medium. They can enter into a replication series and then lie fallow for generations, until someone else happens to stumble upon them to initiate a new series. Or they can be passed on unnoticed, the way that certain atavistic genes are, until they begin to function again. Scientists' brains can serve as vehicles for replication sequences, but scientists themselves are anything but passive vehicles for such sequences. They also function as agents. Without scientists, no conceptual replicator could ever be tested, and testing is essential to science. As a shorthand expression, we frequently talk about the meanings of words and sentences, and in developed languages there is some point to such circumlocutions, but ultimately people are the entities who mean things by what we say. Individual scientists are the agents in scientific change.

In sum, conceptual replication is a matter of ideas giving rise to ideas via physical vehicles, some of which also function as interactors. Replicators are generated, recombined, and tested by scientists interacting with the relevant portion of the natural world. Because I see a ball accelerate as it rolls down an inclined plane, I come to hold beliefs about the motion of balls as they roll down inclined planes. Something in the non-conceptual world initiated a replication sequence in the conceptual world. These sequences of events in the non-conceptual world are the sorts of causal connections that natural science is designed to discover. Social scientists study the perceptual connections between individual organisms and the rest of the world, including other organisms.

Causal connections exist between scientists and the non-social natural world, but they also exist among scientists themselves. Science is not only a conversation between individual scientists and the natural world but also a conversation among scientists. Causal connections also exist between scientists and their societies at large. For example, throughout its history, science has been carried out in sexist societies. Perhaps this sexism has "infected" science itself. These larger social influences are not always of the sort that need to be "overcome," but when they are, the social organization of science permits it. Because the career interests of individual scientists frequently conflict with each other while coinciding often enough with the manifest goals of science, scientists not only come to notice the effects of broader social interests on their belief systems but also sometimes overcome these effects. Perhaps scientists raised in a sexist society should not be able to notice the sexism latent in their society let alone overcome it, but they do.

Previously I noted that replicators must not only exhibit structure but also pass it on to subsequent replicators. But structure alone is not enough. This structure must count as "information." In the case of selection, biological evolution has been taken to be the literal usage, and the transfer to conceptual change is usually considered, if not dismissed as being, "merely" analogical. In the case of information, conceptual transmission is the literal usage, and the genetics context is analogical. Many regularities exist in nature. Planets travel in ellipses, gases expand when heated, molecules are transported differentially through semipermeable membranes, and crystals form very regular shapes. None of these regularities count as "information" in the appropriate sense. The order of bases in molecules of DNA does because of the character of this order, its origins, and its effects.

Much of the structure of DNA is strictly lawful. Given certain general constraints, it has to be the way it is. For example, if the external "backbones" of a DNA molecule are to be kept equidistant down its length, then the "rungs" between these backbones must be kept of equal length. As its name implies, deoxyribonucleic acid is an acid and, as such, must have the general characteristics of acids. These are the features of DNA which allowed scientists to unravel its general structure. The order of bases in a particular molecule of DNA is quite different. Adenine can bind only with thymine and quanine can bind only with cytosine, but with only minor exceptions, any of the four bases can precede or follow any of the other bases as well as itself. All orders are equally likely from the perspective of physical law. That is why the order that happens to exist can function as a code. With equally minor exceptions, any letter in a language such as English can precede or follow any other letter.

As problematic as the distinction is between those features of a molecule that follow lawfully from the fundamental character of the physical world and those that are contingent, it is nevertheless important. The distinction between a code (or language) and particular messages expressed in that code is similarly quite important. Both natural languages and messages are built up historically and can be used to infer history. Numerous genetic codes were possible when life evolved on Earth. Which code happened to evolve is primarily a function of the particular circumstances that obtained at its origin. Because all terrestrial organisms use the same code with exceptions that are as slight as they are rare, the assumption is that all life here on Earth had a single origin. Although the genetic code is relatively simple when compared to a natural language such as English, it is still sufficiently complicated that the likelihood that exactly the same code could have evolved twice is extremely small. Although all organisms use the same code, quite obviously they contain different messages. Because later messages are modifications of earlier messages, molecular biologists have been quite successful in reconstructing the past history of extant forms of life.

In a sense, any record of the past can serve as information about the past. For example, unusually high concentrations of iridium in thin layers of the geological strata can be used to infer past impacts with meteors, footprints of dinosaurs in the hardened mud of a river bed imply the existence of these organisms at the time that this layer was formed, and masses of charred wood indicate a forest fire. Organisms can learn about the world in which they live either directly by interacting with it or indirectly by observing some other organism do the interacting. Learning from experience in the first instance may initiate a replication sequence but does not itself count as replication. In their investigations, scientists learn about the structure of the empirical world. They record this knowledge in a language of some sort. This characterization of the natural regularity counts as information, but the natural regularity itself cannot without making the notion of information vacuous. The chief exception is knowledge of the genetic code. In this instance, the regularity that initiates a conceptual replication sequence is itself part of a replication sequence.

Scientists have carried the process of learning to its extreme. Students can run experiments themselves or watch their instructors do so. But most learning in science comes from reading or hearing about the activities of others. Only a small portion of what a scientist believes about the world arises by means of this scientist interacting with the relevant phenomena. Each scientist has only a few decades to contribute to science. Time cannot be wasted checking every knowledge claim before using it. Using without testing makes scientific progress possible, but it also increases the possibility that some of these knowledge claims are likely to be mistaken. However, one should not forget that knowledge by acquaintance, no matter how direct, is also far from infallible.

DISANALOGIES BETWEEN BIOLOGICAL AND CONCEPTUAL EVOLUTION

Numerous differences have been alleged between biological and conceptual change. Most have very little substance and can be treated quite briefly. Others have some point and deserve a fuller discussion than I can give here. For example, one frequently hears that conceptual evolution occurs much more quickly than biological evolution. In point of fact, conceptual evolution occurs at an intermediate rate as far as physical time is concerned. Viruses evolve much more quickly than conceptual systems in even the most active areas of research, while large, multicellular organisms evolve more slowly. However, physical time is relevant only to interaction. As far as replication is concerned, the relevant metric is generation time. With respect to generations, conceptual evolution occurs at the same rate as biological evolution – by definition.

Some authors argue that no general analysis of selection processes equally applicable to biological and conceptual evolution is possible because genes are "particulate" while the units in conceptual replication are highly variable and far from discrete. In point of fact, neither biological nor conceptual replicators are all that "particulate." In both cases, the relative "size" of the entities that function either as replicators or as interactors is highly variable and their boundaries sometimes quite fuzzy. If the entities that function in selection processes must all be of the same size and/or be sharply distinguishable from each other, then selection can no more occur in biological than in conceptual contexts.

Another objection that has been raised is that biological evolution is always biparental, while conceptual evolution is usually multiparental. Once again, this objection is based on a simple factual error. For a large number of organisms, inheritance is biparental; for most it is not. In conceptual evolution, rational agents sometimes combine ideas from only two sources; sometimes from several. Upon casual inspection, polyploidy seems somewhat more common in conceptual than in biological evolution, but that is all. When one switches from the level of individual entities to populations, no significant differences can be found between the two. At a particular locus, numerous different alleles can coexist in various frequencies. Numerous different solutions to the same problem or versions of the same idea can coexist in conceptual populations.

Upon first glance, cross-lineage borrowing seems much more common in conceptual than in biological evolution. If lineages are defined in terms of replication sequences, extensive cross-lineage borrowing is ruled out by definition. Two lineages can remain distinct in the face of some cross-lineage borrowing, but once it becomes too extensive, the two lineages merge into one. Regardless of common-sense beliefs, gene exchange does occur between groups that are considered different species, and the amounts of gene exchange needed to neutralize any genetic differences between two largely disjoint lineages turn out to be quite small. In short, in biological evolution, extensive cross-lineage borrowing cannot occur because lineages are generated by this very process. When conceptual and social lineages are distinguished in science, extensive cross-lineage borrowing becomes possible, i.e., scientists belonging to different socially-defined groups can and sometimes do use each other's work. In such situations, the groups remain socially distinct while their conceptual correlates merge. However, such cross-lineage borrowing in science does not seem to be as extensive as recurrent references to various "syntheses" would lead one to expect. Rarely do conceptual lineages merge without the scientific communities that produced them merging as well. Mergers of both sorts occur in science. They occur in biology as well, especially among plants. So far no one has produced the data necessary to see in which contexts cross-lineage borrowing is more prevalent.

The most commonly cited disanalogy between biological and conceptual evolution is that biological evolution is Darwinian while conceptual evolution is largely Lamarckian. No organism is able to pass on any of the ordinary phenotypic traits that it acquired during the course of its existence to its progeny, but some organisms can pass on what they have learned about their environment through social learning. As often as such observations are repeated, no one goes on to explain at any length what they mean. No one claims that conceptual evolution in science is *literally* Lamarckian, as if the basic axioms of quantum theory are somehow going to find their way into our genetic make-up. If conceptual entities are taken to be phenotypic traits, then conceptual evolution is not literally Lamarckian because changes in these traits leave genes untouched. Ideas are transmitted but not inherited. If simple transmission is sufficient for Lamarckian inheritance, then a mother giving her baby fleas counts as Lamarckian inheritance. Taken *metaphorically*, conceptual evolution is still not Lamarckian because ideas (or memes) are held to be the analogs of genes, not characters. If anything, conceptual evolution is an instance of the inheritance of acquired memes, not characters. We do learn from experience and pass on this knowledge socially, but I fail to see why these processes should be considered "Lamarckian" in either a literal or a metaphorical sense. On the literal interpretation, ideas count as acquired characters, but the transmission is not genetic. On the metaphorical usage, ideas count as analogs to genes, not characters. Although the genotype-phenotype distinction can be made in the context of conceptual change, the net effect is that the analogs to phenotypes are not inherited. In the absence of anything like the inheritance of acquired characters, I think that characterizing conceptual change as "Lamarckian" leads to nothing but confusion.

As far as I can see, the *only* sense in which conceptual evolution is Lamarckian is in the most caricatured sense of this much-abused term, i.e., it is intentional. Just as giraffes increased the length of their necks by striving to reach leaves at the tops of trees, scientists solve problems by trying to solve them. Science is intentional, in fact it is as intentional as any human activity can get. We learn about the natural world by contriving to interact with it. For some, the gulf that separates intentional acts from the rest of nature is so wide and deep that no comparisons are possible. I do not share this conviction, but I have no in-principle arguments that are liable to touch those who wish to insulate the behavior of intentional agents from the sort of principles that apply to the rest of the natural world. All I can do is to point out some of the consequences of taking this distinction as primary and all others as secondary. For example, in the *Origin of Species*, Darwin reasoned from the known effects of artificial selection to the possible effects of natural selection. But artificial selection is intentional. Perhaps plant and animal breeders cannot produce mutations at will, but they do consciously choose those organisms to breed that exhibit traits which they find desirable. If reasoning from artificial selection to natural selection is totally illicit, then Darwin's main argument in the *Origin* is one gigantic blunder. Similarly, any extension from controlled experiments in science to the rest of nature is illicit.

I also do not think that the role of intentionality in scientific contexts is actually at the root of what bothers critics about any attempt to provide a single analysis of "selection" that applies equally to biological and conceptual evolution. Scientists do strive to solve problems. They both generate novel ideas and select among them. Right now genetic mutations occur by "chance." However, in the very near future, biologists will be able to generate any genetic mutations they see fit. When that occurs, intentionality will play the same role in both biological and conceptual change. I doubt that in such an event critics will instantly become converted. If I am correct in my guess, then the role of intentionality in generating novelty must not have been all that important of an objection in the first place.

For my part, I think a better way of classifying selection processes is between, first, those that are gene-based and those that are meme-based and only then worry about the complications which ensue from intentionality. Some gene-based selection is intentional; most is not. Some meme-based selection is intentional; most is not. On this classi-fication, artificial and natural selection are fundamentally the same sort of phenomenon. On this same classification, the sort of rational selec-tion of beliefs in which people engage (when all else fails) and all the semiconscious and unconscious selective retention that characterizes how human beings acquire their beliefs are also fundamentally the same sort of phenomenon. Once one looks at science as a whole, paying attention not just to the rare scientist who happens to make a major contribution to science but to the vast army of scientists who have no discernible impact on science, the effects of intentionality do not look so massive. If scientists did not strive to solve problems, the frequency with which they succeed would no doubt decrease, but it is already so low that the differences would be difficult to discern. All scientists are constantly striving to solve problems. Few do. Of those who do, only a very few are noticed. There may well be a difference in kind between intentional and non-intentional behavior, but it is not a difference in kind that results in much of a difference in degree.

One difference between biological and conceptual evolution is that in biology genes make genes. In the most primitive circumstances, genes were also probably the only interactors. Eventually, however, they began to produce more inclusive entities that could promote repli-cation by interacting with their more inclusive environments. Concep-tual replicators do not, on their own, produce copies of themselves. They do so only via their most important agents – individual scientists. Thus, in scientific change, scientists are the chief agents in both repli-cation and interaction. However, on my analysis, this difference is not sufficient to preclude a single analysis applying equally to both.

Another apparent difference between biological evolution and con-ceptual change is that biological evolution is not clearly progressive while in certain areas, conceptual change gives every appearance of being progressive. At a glance, biological evolution appears to be as clearly progressive as conceptual evolution in the most advanced areas of science, but appearances are deceptive. Thus far biologists have found it surprisingly difficult both to document any sort of biological progress in the fossil record and to explain what it is about the evolutionary process that might lead phylogenetic change to be progressive (for the most convincing data to date, see Signor 1985).

Conceptual development in certain areas of human endeavor, especially in certain areas of science, gives even a stronger appearance of being progressive. Although science is not progressive in the straightforward way that earlier enthusiasts have claimed, sometimes later theories are better than earlier theories even on the criteria used by advocates of the earlier theories. Science at least appears to be more clearly progressive than biological evolution. Of greater importance, we have good reason to expect certain sorts of conceptual change to be progressive.

Intentionality is close to necessary but far from sufficient in making conceptual change in science progressive. It is not absolutely necessary because sometimes scientists have made what turn out to be great advances quite accidentally. Chance certainly favors a prepared mind, but a scientific advance is no less of an advance because the problem which a scientist happens to solve was not the one he or she had intended to solve. The frequency of success in science is quite low. Even so, one should expect that, on average, scientists should solve the problems which they are trying to solve more frequently than those which are only at the periphery of their attention. At the very least, the intentional character of science should speed it up. However, intentionality is far from sufficient in explaining the progressive character of science. If everything about the natural world were in a state of haphazard flux, scientific theories would also continue to change indefinitely, not just because scientists continue to change their minds about nature but because nature itself is changing. Goal-directed behavior can have a direction in a global sense only when the goal stays put.

Whenever the conditions are right, evolution by means of natural selection occurs. The global goal of natural selection may well be increased adaptation, but for particular lineages, the contingencies to which successive generations of organisms must adapt keep changing, not because genetic variation is "blind," not because natural selection is non-intentional, but because so many of the aspects of the environment to which organisms must adapt keep changing. Conceptual evolution, especially in science, is both locally and globally progressive, not simply because scientists are conscious agents, not simply because they are striving to reach both local and global goals, but because these goals exist. If scientists did not strive to formulate laws of nature, they would discover them only by happy accident, but if these eternal, immutable regularities did not exist, any belief a scientist might have that he or she had discovered one would be illusory.

CONCEPTUAL INTERACTION

In biological evolution, replicators pass on their structure largely intact. Some of this structure counts as "information." Through interaction this information is translated into phenotypes at a variety of levels. One reason why selection processes are so complicated is that they evolve via an interplay between two subsidiary processes (replication and interaction) occurring at a variety of levels. Because so little of the information encoded in the relevant replicators is ever realized, selection processes are extremely particularistic and idiosyncratic in their effects. For instance, genes do not literally code for traits. Given any genome, a wide spectrum of traits could eventuate depending on the sequences of environments confronted. Of all the phenomes a particular genome could have produced, only one is produced. A genome that might well have proven to be extremely fit in a wide variety of environments is extinguished because it does not happen to find itself in one of these environments. Even when a single genotype is expressed clonally in numerous genomes, only a tiny fraction of the phenomes that could have been produced actually are produced. Hence, genotypes never get to show "all their stuff." In short, translation entails a tremendous loss of information.

Similar observations hold for conceptual evolution. Scientists do spend an appreciable amount of time testing their views, but as everyone now acknowledges, scientific theories are always grossly underdetermined by anything that might be considered data. Given any set of observations, the number of alternative theoretical explanations that might be generated to account for them is limited only by the ingenuity and good sense of scientists. Of all the possible explanations that could be offered, only a small fraction actually are ever offered. Conversely, of all the observational implications of any one version of a particular theory, only a small percentage actually are ever made, let alone tested. As a result, chance plays a large role in which versions of which theories ever become prominent. A particular theory might generate considerable attention even though it has serious defects because the first observations made happen to be among the relatively few that actually conform to it. Conversely, a theory that in retrospect has much to recommend it might be rejected because the first observations made in testing it happen to be among the relatively few that are at variance with it. The slippage which exists between theories and data is only exaggerated by the number and variety of compromises that must be introduced in order to "operationalize" a theory so that it can be tested.

Biologists were not prepared for the amount of genetic variation that characterizes biological species. More often than not, greater heterogeneity exists within a species than between it and its closest evolutionary congeners. The same heterogeneity characterizes science. Praise for conceptual pluralism in science is currently fashionable. Conceptual pluralism is necessary if science functions as a selection process, but so is conceptual pruning. Taken literally, the maxim that anything goes would be lethal to science. Scientists are constantly generating different combinations of the conceptual tools which have been willed to them by previous generations. At times, in fact quite rarely, genuine novelties are also introduced. At a distance, much of this conceptual heterogeneity is obscured by the deceptive appearance of terminological conformity. Scientists are as terminologically conservative as they are semantically flexible. Under cover of the same terms, scientists working in the same research program frequently hold very different views.

Kuhn (1970) attempted to make his notion of a "paradigm" more operational by reference to scientific communities. Paradigms are those things that members of the same socially-defined scientific community share. The trouble is that if one actually defines a group of scientists in terms of their professionally relevant relations, the resulting groups tend to be conceptually quite heterogeneous. Those of us who study science tend to find this heterogeneity quite disconcerting. We are convinced that the only way that people can cooperate is for them to agree with each other at least over fundamentals. Every scientific theory *must* have an essence, and every scientist working in the same research program *must* accept these essential tenets.

I do not know about people in general, but scientists seem to be able to cooperate with each other even when they are in fundamental disagreement. They do so, in part, by playing down these differences. When asked, scientists insist that the conceptual system that they are developing can be characterized by a set of fundamental propositions about which there is universal agreement. However, when the members of a socially-defined research group list the fundamental principles of their research program, they present different lists. Even when some items on this list are terminologically the same, the intent frequently varies. For example, a group of systematists might all agree that all higher taxa must be monophyletic but mean very different things by "monophyletic."

Much of this heterogeneity is lost as particular research programs "harden." In retrospect a particular program may well appear to have had an essence or Lakatosian "hard core," but as Lakatos himself noted, hard cores can be recognized only in retrospect. While the selection process is going on, conceptual systems are heterogeneous – as they must be if science is a selection process. When scientists say that the research program they are working on has an "essence," they mean the views that they individually happen to hold at the moment. *The essence of a Darwinian view of evolution is their* view, the benighted opinions of others who consider themselves Darwinians notwithstanding. Sooner or later such dogmatism does triumph. One version will come to be accepted as *the* version, and all the variation which characterized the program in its active period will be ignored or discounted. The challenge then is to provide a method for making sense of scientific change in the face of all this conceptual heterogeneity.

THE TYPE-SPECIMEN METHOD OF REFERENCE

On the account of scientific change that I have sketched above, groups of scientists must be distinguished from the conceptual systems that they produce and the two followed separately. Both form internally heterogeneous lineages that can change through time. A research group can persist while old members leave and new members join. Conceptual lineages are no less heterogeneous. At any one time, they can contain contradictory elements, and a particular statement-token can give rise through successive replications to a statement-token that contradicts it. In the face of all this heterogeneity and change, how can those of us who study science make any sense of scientific change? How can we individuate and refer unambiguously to a particular lineage, either social or conceptual? One possible solution to these problems is the appropriation of a method devised by systematists through the centuries to handle parallel problems with respect to biological species.

When systematists come across what they take to be a previously unknown species, they pick a specimen, any specimen, and designate it as the type specimen. In spite of the connotations of this term, the type specimen need not turn out to be in any sense "typical." All a type specimen does is determine to which species a name applies. No matter how aberrant a type specimen may turn out to be, it belongs to a particular species and to no other. It is one node in the genealogical nexus. No matter how the boundaries of a particular species are reworked, the species that includes the type specimen must be called by the name that the type specimen bears. Both the name and the type specimen are passed down through the generations from systematist to systematist.

As similar as the type-specimen method in systematics may look to the theory of rigid designation (Kripke 1972, Putnam 1973), it differs from it in several important respects. In systematics fictitious baptisms play no role, and systematists treat link-to-link transmission chains seriously. Systematists actually perform the sort of historical inquiry necessary to trace names back to their sources and make their decisions accordingly. Priority is one of the key elements in their codes of nomenclature. But most importantly, biological species are chunks of the genealogical nexus and as such count as spatiotemporal particulars and not classes or kinds. As a result, their names are best construed as being proper, not general. Because proper names have been treated traditionally as "rigid," one need introduce no new theory of reference to accommodate them. Like an organism such as Moses, a species such as *Dodo ineptus* has a beginning, middle, and end. A name can be attached to it rigidly during any time-slice of its existence. If one chooses, one can apply the same name to a lineage throughout its existence, no matter how much it might change, or subdivide it, giving each subdivision a separate name. For example, one might choose to subdivide an organism into sequential stages and name each, just as one can subdivide a gradually evolving lineage into successive chronospecies, giving each a different name, but the logic of the situation remains the same. The lineage is basic; the characters describing the entities that are part of the lineage are secondary.

The type-specimen method works so well for historical entities because *both* the entity being named *and* the subsequent link-to-link transmission of its name form historical entities that can be traced independently of meaning change to see if, in the past, they intersect in the way claimed. Did someone named "Moses" exist at the time and place claimed in the Bible? Did the dodo ever exist on Madagascar and only recently go extinct? These questions have answers which are independent of a whole variety of other considerations. "Moses" would still designate Moses even if he did not do many of the things attributed to him in the Bible, and "dodo" would still designate the dodo even if we happily discovered a population of these birds still alive in some remote valley somewhere.

The type specimen does not work quite so well for terms that are genuinely general because the entities referred to are not themselves historical entities. The substances gold and water can exist anywhere in the universe whenever the conditions are right. They can be named numerous times both here on Earth and elsewhere. Because languages here on Earth have histories, and these histories are to some extent interconnected, sometimes link-to-link transmissions of term-tokens can be traced back through time, and sometimes they converge, but they need not. "Wasser" and "water" may well have the same terminological ancestor, but the ultimate reason why the plethora of terms for denoting this substance are held to denote the same substance is similarity in structure of this substance.

However, if conceptual change in science is taken seriously as a selection process, then something like the type-specimen method can handle it, but this revised theory depends crucially on organizing term-tokens into term-trees solely on the basis of transmission. In these trees, the structure of the tokens can change, e.g., tokens of "pangen" can be transcribed as "pangene" and then "gene." Even so, these term-tokens belong to the same token-tree. Even the character of the events that initiate the use of term-tokens can change. For example, De Vries used tokens of the term "pangen" only in certain circumstances, circumstances significantly different from those that elicited the tokens of the term "gene" from Johannsen. Present-day scientists use this same term under an even greater variety of situations. What binds all these term-tokens together is that they all belong to the same term-token tree. Actual baptisms did occur, and subsequent uses actually form intersecting trees (for further discussion of the application of the type-specimen method to conceptual change in science, see Hull 1983b; for objections, see Mayr 1983).

One can group organisms together in a variety of ways. Which ways are preferable depend on the use that can be made of these groupings.

For example, one can group organisms into those that reproduce sexually and those that do not. If these groups are treated as being genuinely general, then the terms referring to them might well function in laws of nature. Or one might group organisms according to descent. If so, then these groups must be spatiotemporally restricted, and the terms that denote them cannot function in spatiotemporally unrestricted generalizations. However, at the very least, grouping by descent serves to recognize the entities that result from natural selection. Some of these groups themselves might even function in selection processes. The distinction is between "Carnivora" and "carnivorous." The former refers to a chunk of the genealogical nexus; the latter to a group of organisms that share the ability to eat and digest meat. The two groups are far from coextensive. Not all species that belong to Carnivora are carnivorous, and many species of carnivorous organisms do not belong to Carnivora. Both sorts of groupings have their function in biology, but these are different functions.

But, one might object, why not equivocate between these two different uses of the term? Certainly that is what most people do most of the time in ordinary language. Sometimes "Baroque" is used as if it applied just to a particular period in human history, sometimes in a more general sense. Why not term a chair or building constructed today "Baroque"? Insisting that this term applies only to a particular time and place is being too "monistic." Although distinguishing between Tiffany lamps and Tiffany-type lamps might seem overly pedantic, it is no more pedantic than distinguishing between characters such as eyes (in the sense of any organ that can be used to discern light) and eyes as evolutionary homologies. Vertebrate eyes and cephalopod eyes are not the "same" character. In the context of selection processes, these two senses of "same" must be distinguished. Some genes are identical, some identical by descent, and the roles in population genetics of these two sorts of identity differ (Hull 1986).

Parallel distinctions exist for term-tokens. They can be grouped into types solely on the basis of something such as "similar meanings" regardless of genesis. If so, than any token of this term-type that has the appropriate meaning belongs to that type. These term-tokens can appear anywhere at any time. Although it is unlikely that the same term-type should be coined independently to refer to the same thing, it would make no difference if such an unlikely occurrence did take place. In fact, if each term-token of a term-type were generated *de novo* with each utterance, it would make no difference. The notion of "same-ness of meaning" has proven to be extremely slippery. Even so, if same-ness of meaning is what groups term-tokens into term-types, then genesis is irrelevant.

If one wants to treat conceptual change as a selection process, then term-tokens must be grouped into lineages and trees by means of trans-mission. Link-to-link transmission must be taken literally. In these replication sequences, term-tokens themselves can change (e.g., "pangen" can become "pangene"). They can also change the means by which they are connected to their referents (e.g., additional operational "definitions" for "gene" can be introduced). As strange as this way of grouping term-tokens may seem, it is necessary if conceptual systems are to evolve by means of selection.

Both ways of grouping term-tokens have their function in science. Within the context of a particular controversy in science, the causal connection of term-tokens in replication sequences is crucial. Term-tokens are the things that are being differentially perpetuated. Anyone who wants to understand scientific change at the local level must order term-tokens into trees. As "irrational" as it may seem, scientists evalu-ate claims in terms of their genesis because of the influence of both conceptual inclusive fitness and the demic structure of science. Two instances of the same statement-type are evaluated differently if they happen to be part of two different lineages. One might be rejected, resulting in that conceptual lineage going extinct. The other might be accepted and proliferate until it is universally accepted. However, those involved in these selection processes intend their usage to be general. They intend to transmit term-types, but all they actually transmit are term-tokens which are immediately interpreted as types.

Thus, conceptual change in science viewed as a selection process incorporates a systematic ontological equivocation. Term-tokens are tested and transmitted locally but interpreted globally as types. Term-tokens are simultaneously part of spatiotemporally restricted term-trees and instances of spatiotemporally unrestricted term-types. Each generation of scientists intends for their conceptual systems to be gen-erally applicable and universally accepted, but in each generation only a very small percentage of instances of these systems gets passed on, and the version of a particular conceptual system that eventually comes to prevail may well not be the one that early scientists intended. As strange as the distinction between identity and identity by descent may appear, it permeates all evolutionary biology. For many processes, iden-tity by descent is required; for others it is not. Similar genes or traits behave similarly in similar situations, differences in genesis notwithstanding. If bottlenecks are as important in biological evolution as many evolutionary biologists claim they are, then identity by descent is crucial in the evolutionary process. Most change occurs when a population goes through such a bottleneck. Similarly, if small research groups are as important as some students of science claim that they are, then similar observations hold for the role of identity by descent in conceptual change in science.

CONCLUSION

Every explanation takes certain things for granted and explains other things in terms of them. I have taken for granted that scientists are by and large curious about the world in which they live and desire credit for their contributions to science. I provide no explanation for these characteristics of scientists. The human species seems innately inquisitive. The process by which young people are introduced into science at least does not totally destroy this native curiosity. In some cases, it even encourages it. According to Harré (1979), the desire for recognition from one's peers is equally strong in human beings. Even if a budding young scientist enters science not caring about something as paltry as individual credit, he or she will find it very difficult not to get caught up in the general enthusiasm. I have also not presented any justification for our belief in an external world which we can come to know or for the existence of any regularities in nature. However, given curiosity, a desire for credit, and the possibility of checking, the structure that I claim characterizes science can explain quite a bit about the way in which scientists behave.

Many commentators find one or more features of science as it has existed for the past couple hundred years to be less than palatable. They find scientists too polemical, aggressive, arrogant, and elitist. Scientists are too anxious to publish so that they can scoop their competitors. They seem more intent on enhancing their own reputations than in helping humanity. According to these commentators, scientists should turn their attention from the problems that they find most interesting to those that are currently most relevant to our survival. The mechanism which I propose explains why scientists do not behave in the way that these critics think that they should behave. From an operational point of view, behavioral psychologists are right on at least one count: organisms tend to do what they are rewarded for doing, pious hortatory harangues notwithstanding. If scientists are rewarded for making new discoveries, formulating more powerful theories, designing novel experiments, etc., then they are likely to do just that. Perhaps scientists could be raised so that they were not so strongly motivated by curiosity and the desire for individual credit, but I am not sure that the results would be worth the effort. In fact, such efforts, if successful, might bring science to a halt. At the very least, in the absence of the mechanism which I have sketched, science could be likely to proceed at a very leisurely pace.

The mechanism that I have sketched in this paper may not seem like much of a mechanism, hardly up to explaining the marvelous progress made during the past few centuries by successive generations of scientists. But when one thinks of it, natural selection is not much of a mechanism either, and yet it has produced all the fantastic adaptations that organisms, both extinct and extant, exhibit. The mechanism that I sketch is also not very efficient. If nearly all the progress in science turns on the work of a very small percentage of scientists working at any one time, then science could be made much less expensive and no less efficient by weeding out those scientists who are not being very effective. However, if biological evolution has any lessons to teach us, it is that selection processes cannot be made too efficient without neutralizing their effects. The sort of interindividual and interdemic polemics that have characterized science from the beginning are not a very efficient way of reconciling differences among competing scientists and groups of scientists, but they are extremely effective.

NOTES

Reprinted from *Biology and Philosophy* 3 (1988): 123–55 with kind permission from Kluwer Academic Publishers.

I wish to thank both Michael Ruse and Ronald Giere for suggesting improvements in an early draft of this paper, which is an abstract of *Science as a Process* (1988a). The number of people who helped me in developing the ideas set out in this book was extremely large, so large that I decided to defer expressing my gratitude to them until its appearance.
1. In addition to omitting data from this paper, I have kept references to a minimum. Both of these omissions are serious, given the mechanism for conceptual change that I set out in this paper.
2. From past experience, I have learned that the preceding distinctions are difficult to keep straight, in part because ordinary English is not constructed to make them. I am arguing that the species category might well be a natural

kind with a central role in biological evolution, that *Homo sapiens* might well be an instance of such a natural kind, but that the human species itself is not a natural kind. Parallel distinctions with respect to a physical element such as gold are as follows. Physical element is a natural kind. One instance of a physical element is gold. It too is a natural kind. However, the pope's ring is only an instance of gold.

3. "Vehicle" is used in two distinct senses in the literature on selection processes. Campbell (1979) uses it as I do to refer to the physical entities that are functioning as replicators. Dawkins (1976) uses it to refer to interactors. Williams (1985) prefers not to term organisms "vehicles" because it plays down the active role of organisms in biological evolution. Needless to say, I prefer Williams's usage.

5

Why Scientists Behave Scientifically

In the midst of all the debunking of science that is currently fashionable, we tend to lose sight of the fact that science has been and continues to be more successful than any other social institution in fulfilling its stated goals. Of course, science does not have to work all that well to be more successful than any other social institution (such as Congress) in attaining its stated goals. But much more can be said on behalf of science. If the primary goal of science is to increase our knowledge of the natural world, it has been successful beyond anyone's wildest dreams. I admit that very little of this knowledge has found its way into the consciousness of very many human beings. Only a tiny proportion of the human race understands relativity theory, let alone quantum theory, and most people who think that they understand evolutionary theory profoundly misunderstand it. Even so, within its limited domain, science has been extremely successful. The question then becomes, how come?

In 1953 Charles E. Wilson, the President of General Motors, became famous for saying that what is good for General Motors is good for the country.[1] His contemporaries can be excused for being a bit suspicious of such a self-serving claim. But science is fortunate that science is so organized that, by and large, what is good for the individual scientist is good for science. Scientists want credit for their contributions. They want other scientists not only to notice their work but also to use it, preferably with a few generous citations. One of the peculiarities of science is that the first person who publishes a view (or more accurately gets the earliest submission date) gets all the credit, even though several other scientists may have been almost there. As incredible as it may seem, the winner-takes-all convention in science arose in the 17th century in order to force scientists to publish. *Force* scientists to publish! Scientists would like to keep their discoveries under wraps long enough to milk them for their most obvious consequences, but if they withhold them from publication for too long, they are likely to get scooped and get no credit at all. The legacy of this early convention today is an unseemly rush to publish.

However, a system of citation has also arisen that holds in check this rush to be the first to make a view public. Scientists cite the work of other scientists in part to give credit where credit is due but also in part to gain support for their own views. Thus, scientists are caught in a bind. They want their work to be accepted, but they also want it to appear as original as possible. Showing that it flows naturally from the well-established work of one's contemporaries is likely to increase the likelihood that it will be accepted, but such a practice automatically detracts from its originality. Conversely, omitting any references to the work of others makes one's own contributions look highly original but also decreases the likelihood that one's fellow scientists will take it seriously enough to incorporate it into their own work. In general, if a scientist is sparing in citing the work of others, these other scientists are likely to return the compliment. In short, scientists trade credit for support and vice versa. For each opportunity to cite, a scientist can have one or the other, but not both.

Of course, mutual citation can be found in a variety of professional institutions, but what distinguishes science from other professions in this regard is that scientists operate with a notion of truth that is much easier to apply within science than outside it. In fact, one of the defining characteristics of science is the ability to test one's views about the natural world in a reasonably direct way. Although scientists do not test each other's results as often as some naive commentators seem to think that they should, replication does occur in science. One of the strengths of science is that not all results need to be tested. Scientists amass lots of data, some of it fairly isolated, but they also devise theories which organize data and entail all sorts of conclusions about what should be the case. Any error fed into the system is very likely to produce erroneous results elsewhere.

If scientists had to check each and every result before they incorporated it into their own work, science would slow to a crawl. Instead, scientists tend to trust the results produced by others. However, not every scientist engenders the same amount of confidence. Some have the reputation of publishing work that is too fast and dirty, while others produce results that may not be all that exciting but at least you can depend on them. Scientists are constantly enjoined to adhere to the strictest canons of good scientific practice. Such invocations may have some positive effect on how scientists behave, but in other areas of human endeavor, comparable calls to do one's duty hardly seem sufficient to bring about the stated goals of the discipline. For example, physicians who own their own CAT scan machines find that their patients need double the number of such procedures as do physicians who have no financial interest in such machines.

Calls to do one's duty to the larger group certainly have some effect, but it always helps if individuals do not have to sacrifice their individual goals for the good of the group. Social systems work much better when virtue and self-interest go hand in hand. Once again scientists are in a bind. They would like to conduct their research as quickly as possible, to get their results out there sooner than anyone else so that they can get the credit for their discoveries. But the chief credit in science, the currency that really matters, is *use*. Scientists use each other's results, almost always without testing them. However, if something starts going wrong with their own research, scientists begin searching to see what went wrong and why. If the error can be traced back to your work, you are in real trouble. Citations may well give credit where credit is due, but they also leave paper trails for assigning blame as well. With the possibility of credit comes the possibility of *dis*credit.

Because scientists are invested in their own work, they are not all that good at discovering errors in their own pet hypotheses, but other scientists are more than happy to fill the gap. If anything, the sort of testing that goes on in science can be too rigorous. Scientists get very little credit for replicating other scientists' experiments, but they do get credit for discovering mistakes in the work of others, especially if this research is taking place in one of the "hot" areas of science. The rush to publish, when properly constrained, increases the pace of science. The monitor on this pace is the punishment meted out to those scientists who produce unreliable work. Some errors are more understandable, more excusable than others, but any error impedes the research of anyone who uses it. Failure to include appropriate citations hurts the careers of the scientists who are not cited. Erroneous results hurt the careers of everyone who uses them, and they are very likely to retaliate.

Thus, science can be viewed as a self-policing system of mutual exploitation – or cooperation if one prefers. It works only when individual and group interest coincide. As scientists are increasingly able to make money off their discoveries, the same sorts of financial impropriety that characterizes all other professions will increasingly characterize science.[2] Whenever scientists serve two masters, compromises will be made, whether these masters are government, industry, or mammon. Throughout most of the history of modern science, scientists have behaved extremely well as far as determining truth is concerned,[3] not because scientists are inherently superior beings, but because it has been in their own best self-interest to do so. Many scientists may be excellent candidates for sainthood, but one reason why science has worked so well is that scientists need not be saints to contribute to it. As the fathers of our country noted, the "best security for the fidelity of mankind, is to make interest coincide with duty."[4]

NOTES

Reprinted with permission from *MRS Bulletin*, May 1996, p. 72.
1. For a fuller discussion of this quotation, see Dennett (1995, p. 324).
2. Grinnell (1993).
3. As the frequency and bitterness with which priority disputes are fought amply shows, scientists are not so virtuous when it comes to assigning credit. Scientists are also not quite as concerned with the good of humankind as numerous public declarations would have us believe.
4. Hamilton, Madison, and Jay (1788, p. 452).

6

What's Wrong with Invisible-Hand Explanations?

In the midst of all the debunking of science that is currently fashionable, we tend to lose sight of the fact that science has been and continues to be more successful than any other social institution in attaining its stated goals. That some critics wish that science had goals different from the ones that it has is another matter. Of course, science does not have to work all that well to be more successful than any other social institution in attaining its stated goals. It does not take much to work better than our legal system, Congress, or academia for that matter. At worst, science is the tallest midget in the freak show.

A few years back, I published a book (Hull 1988a) in which I argued that part of the explanation for science being so successful can be found in its social organization, and the most important part of this social organization is a mechanism that is commonly termed "invisible hand." In utilizing invisible-hand explanations, I did not reason from economics to the social structure of science, nor did I attempt to gain support for my views by alluding to economics. I was aware at the time that invisible-hand explanations had somewhat of a tarnished reputation in the social sciences, but I did not realize how tarnished until I read the reviews of my book. For example, Miriam Solomon (1995, p. 294) characterized invisible-hand views of science such as mine "optimistic fantasies" (see also Solomon 1996). Other participants in this symposium add to these criticisms, especially subsequent attempts to model such processes in a semi-quantitative way (for an excellent summary of my views as well as socially concerned criticisms, see Ylikoski 1995).

Because of these criticisms I have been led to look more closely at invisible-hand explanations. What are invisible-hand explanations, and what is wrong with them? In this paper I begin by presenting a general analysis of invisible-hand explanations, then set out the invisible-hand mechanism that functions in science, and finally see how this invisible-hand mechanism fits my general analysis.

2 INVISIBLE-HAND EXPLANATIONS IN THEIR CAUSAL CONTEXT

In the philosophical literature, invisible-hand explanations are a special sort of explanation in terms of unintended consequences. Individuals behave the way that they do in order to bring about one goal and in the process bring about some other unintended consequence as well. In the typical case, the intended goals are calculated to benefit the individual, while the unintended consequences have a more general effect. Sometimes the unintended consequences are good, sometimes bad. In Adam Smith's (1776, pp. 129–30) classic example, agents pursuing their own individual gain result in the realization of an unintended benefit for themselves and others. Conversely, the tragedy of the commons is an example of an "invisible-hand" explanation that has bad consequences. Although none of the farmers using the commons intend to destroy it, that is what they do (Hardin 1977). If invisible-hand explanations are optimistic fantasies, then apparently the tragedy of the commons is a pessimistic fantasy, though no one has yet to call it that, possibly because the phenomenon is all too apparent.

However, in the economic literature, the distinction between intentional and non-intentional behavior is of secondary importance. What really matters is that the invisible hand must move a system toward equilibrium. If invisible-hand explanations are limited only to those systems that are moving toward or are at equilibrium, then they certainly do not apply very well to the course of science. One of the most important features of science is that it changes and, it is hoped, will continue to change. Particular theories may well come to map a particular sort of natural phenomena more and more accurately, but such periods are occasionally interrupted by spasms of fundamental change that restructure that area of science from the bottom up. The history of science does not look much like a movement toward equilibrium.

In addition, Mirowski (1994, p. 566) complains that invisible hands are introduced primarily to compensate for treating scientists as if they were marooned Robinson Crusoes, when in actuality scientists form research groups of varying degrees of integration. If economic theories lack any significant idea of "the social," then economics is in real trouble. But invisible-hand explanations need not treat scientists as hermits. All that is necessary is that sometimes these individuals act in relative autonomy from each other with respect to the general good at issue and that, within the groups that they form, an important concern is looking out for themselves.

Several authors have also noted that invisible-hand explanations often result from differences between appearance and reality. A phenomenon that appears to have been brought about by someone's intention actually was not. It was an unintended consequence of human intentions. Conversely, sometimes what appears *not* to have been brought about by anyone's intention actually was. How could all the electricity on the east coast of America go out as it did a few years back? Either some evil cabal was responsible or else a whole series of flukes produced the shutdown. People who have a much higher opinion of the foresight, understanding and competence of their fellow human beings than I have are prone to such conspiracy theories. I myself tend to prefer the second alternative. In any case, Nozick (1974, p. 19) terms conspiracy-theory explanations "hidden-hand" in contrast to "invisible-hand" explanations. In this paper, I limit myself to invisible-hand explanations.

The real problem with invisible-hand explanations is the specification of the mechanism that is supposed to bring about the result (Pettit 1996, Ylikoski 1995). As Ullmann-Margalit (1978, pp. 267–8) concludes, the onus of invisible-hand explanations "lies on the process, or mechanism, that aggregates the dispersed individual actions into the patterned outcome: it is the degree to which this mechanism is explicit, complex, sophisticated – and, indeed, in a sense unexpected – that determines the success and interest of the invisible-hand explanation." Can the mechanisms responsible for this behavior in science be set out explicitly? Are these mechanisms adequate to bring about the effects I claim for science, or do they merely appeal to mysterious coincidences that serve only to paper over these anomalous states of affairs (Geertz 1973, p. 206)?

In sum, invisible-hand explanations are offered in human affairs when individuals behave the way that they do in order to bring about their own self-centered goals and in the process bring about some other unintended more general consequence for a larger group of people of which they are part. In the remainder of this paper I set out what I take to be the mechanisms that are responsible for the peculiar epistemic character of science and argue that these conditions are frequently realized in science as it has been practiced for the past couple hundred years in the West. I begin with some uncontroversial observations about science and scientists – not about *all* scientists *all* the time but about scientists when they are being the *most successful* in coming to understand the world in which we live.

3 A SOCIAL STRUCTURE OF SCIENCE

As science dribbled into existence, it rapidly became a winner-take-all market (Frank and Cook 1995). In order to force scientists to make their achievements public so that other scientists could use them, the convention developed that the first to make a discovery public gets all the credit. In addition, science (like basketball and ballet) is a high-resolution activity. Most of the credit goes to a small percentage of scientists, and scientists *do* want credit for their contributions. They want other scientists not only to notice their work but also to *use* it, preferably with a few generous citations. Scientists cite the work of other scientists in part to give credit where credit is due but also in part to lend support for their own views.

Thus, scientists are caught in a bind. They want their work to be accepted, but they also want it to appear as original as possible. Showing that it flows naturally from the well-established work of one's contemporaries is likely to increase the likelihood that it will be accepted, but such a practice automatically detracts from its originality. Conversely, omitting any reference to the work of others makes one's own work look more original but also decreases the likelihood that one's fellow scientists will take it seriously enough to incorporate it into their own work. Even if they do, they are likely to be as generous with their citations as you were with yours.

Of course, mutual citation is far from an infallible guide to either influence or credit, but it can serve as the sort of rough-and-ready operationalization that is so common in science. Some authors seem to expect the study of science to be more precise and conclusive than science itself is. In addition, citations can be found in a variety of professions other than science, including philosophy. The conventions of science go a good deal deeper that just mutual citation. Replication is crucial. Although scientists do not test each other's results as often as some naive commentators seem to think that they should (Collins 1985), replication does occur in science (Parascandola 1995).

In fact, one of the defining characteristics of science is the ability to test the claims made by scientists about the natural world, and one of its strengths is that not all such claims have to be tested. Scientists amass lots of data, some of it fairly isolated, but they also devise theories that organize data and entail all sorts of conclusions about what *should* be the case. Any error fed into the system is likely sooner or later to produce erroneous results somewhere. If scientists had to check each and every result before incorporating it into their own work, science would slow to a crawl. However, as Mirowski has pointed out to me, the social structure of science not only serves to reduce error but also helps scientists to function successfully even in the presence of error.

Scientists are not peculiar in that they are constantly enjoined to adhere to the strictest canons of their discipline. Such invocations may have some positive effect on how scientists behave, but in other areas of human endeavor, comparable calls to do one's duty do not come close to being sufficient to bring about the stated goals of the discipline. Although appeals to duty certainly have some effect, it always helps if individuals do not have to sacrifice their individual goals for the good of the group. Social systems work much better when virtue and individual benefit go hand in hand. Once again, scientists are in a bind. They would like to conduct their research as quickly as possible, to get their results out sooner than anyone else so that they can get the credit. But the chief credit in science, the currency that really matters, is *use*. Scientists do not use the results of other scientists idly. In adopting the views of other scientists or using their data, scientists are voting with their careers (Quillian 1994, p. 437). One of the most important features of *use* as a criterion of *worth* is that only *practicing scientists* can assign it. Nonscientists play many significant roles in science, but only scientists functioning as scientists can use the work of other scientists by incorporating it into their own work.

As mentioned previously, scientists almost never test the results of other scientists before using them. However, if things start going wrong with their own research, scientists begin searching to see what the problem is. If the errors can be traced back to your work, you are in real trouble. Citations may well give credit where credit is due, but they also leave paper trails for assigning blame as well. With the possibility of credit comes the possibility of *dis*credit. One effect of uncovering mistakes in the work of other scientists is increased care in using the work of these scientists in the future. In fact, often one egregious instance is enough for a group of aggrieved scientists to cease using the work of those scientists found guilty.

Because scientists are invested in their own work, they are not all that good at discovering errors in their own pet hypotheses, but other scientists are more than happy to fill the gap. Scientists get very little credit for replicating other scientists' experiments, but they do get credit for discovering mistakes in the work of others, especially if this research is taking place in one of the "hot" areas of science. The rush to publish, when properly constrained, increases the pace of science. The monitor on this pace is the punishment meted out to those scientists who produce unreliable results. Some errors are more understandable than others, but any error impedes the research of anyone who uses it.

Failure to include appropriate citations hurts the careers of the scientists who are not cited. Erroneous results hurt the careers of anyone who uses them. Thus, with respect to error (not fraud!), science can be viewed as a self-policing system of mutual exploitation or, if you prefer, cooperation. It is also the only self-policing profession that actually polices itself to any significant extent. As Ziman (1994, p. 180) remarks, this system is "not just a quaint tribal custom: it is the social mechanism by which reliable scientific knowledge is generated and evolves."

4 INVISIBLE-HAND EXPLANATIONS IN SCIENCE

The first difference between Smith's invisible hand and science as a social activity is that scientists frequently claim to have higher goals in mind. They are not interested in such paltry rewards as citations, research grants and Nobel prizes. They are interested primarily in knowledge for its own sake. I suspect that to a significant extent scientists really do hold such beliefs. As Sir Peter Medawar (1972, p. 87) once remarked, "Scientists, on the whole, are amiable and well-meaning creatures. There must be very few wicked scientists. There are, however, plenty of wicked philosophers, wicked priests, and wicked politicians." And later in connection with the flap over DNA splicing, "Scientists want to do good – and very often do" (Medawar 1977, p. 20).

But philosophers, priests and even an occasional politician also want to do good. Why are scientists better able to realize their higher goals than the rest of us? Are scientists superior beings, as Medawar seems to imply, or might the answer lie in the social organization of science? Part of the answer can be found in the frequent coincidence in science of individual "selfish" goals and the greater good. Because scientists must use each other's results and use implies worth, they are forced to give at least some credit where credit is due. In addition, both truth and falsity ramify. If other scientists make major advances in part because they have used your work, you will receive at least implicit credit. If, to the contrary, the use of your work leads to dead ends or endless error, you will receive discredit, *extensive* discredit, most likely *explicit extensive* discredit.

If scientists were interested exclusively, or even primarily, in the good of science, then priority disputes would be rare or nonexistent, but they are the most frequent source of discord in science, and this discord can hold back the progress of science. In the past decade, two of the most innovative and powerful scientists in the world devoted a significant amount of their time and the time of their co-workers arguing over who really discovered the virus that causes AIDS (Rawling 1994). Millions of people are dying, and these scientists are arguing over priority! Moral indignation to one side, priority disputes are the price that must be paid for the mechanism that emerged to force scientists to make their discoveries public so that other scientists can use them.

The preceding discussion hints at a second difference between the causal mechanisms operative in science and pure invisible hands – scientists do not come close to working in total isolation from each other. A high percentage of scientists work in tightly-knit research teams (Sen 1983), and those scientists working in such groups are much more productive than those working in relative isolation (Blau 1978). From the perspective of reward and punishment, these groups are the relevant "individuals." Even in the absence of such well-integrated groups, scientists influence and are influenced by other scientists, both as individuals and as sources of information. Invisible hands join with invisible colleges. In addition, scientists are much more responsive to local than to global comparisons (Frank 1985). Most scientists do not expect their work to be noticed, let alone used, by all other scientists but primarily by those working in their own restricted area of science. The most important issue is whether the "individual" goals of scientists and the good of science are sufficiently independent so that they can work at cross-purposes. Once again, priority disputes provide some evidence that they can.

As early as 1788 the authors of the *Federalist Papers* can be found observing that the "desire of reward is one of the strongest incentives of human conduct." As a result, the "best security for the fidelity of mankind, is to make interest coincide with duty" (Hamilton, Madison, and Jay 1818, p. 452). Scientists might well be as motivated to produce knowledge for its own sake as they say they are, and perhaps these admirable motivations are sufficient to bring about the production of reliable knowledge, but the really neat thing about the reward system in science is that it is so organized that, by and large, more self-serving motivations tend to have the same effect as more altruistic motivations. Virtue and benefit go hand in hand. Scientists need not be saints to contribute to science. To the extent that scientists are motivated by the high opinion of others as evidenced by the use of each other's work, they will be pressured to behave themselves.

A third peculiarity of invisible-hand mechanisms in science is that scientists can become aware of them (Ylikoski 1995). Uncritical scientists go about their work convinced that they are motivated almost entirely by the quest for knowledge for its own sake, but some have become aware that they may also be influenced by less altruistic motives. What effect might scientists' understanding of how science works have on science? As in so many other cases, might not this self-knowledge endanger the fabric of science? Perhaps a less visible hand might produce still better science, as Fuller (1994, p. 601) suggests? I think not. As scientists come to understand the various tensions and conflicts that confront them in science, my best guess is that they will be better able to handle them (Hirshleifer 1977).

The preceding may sound plausible, but is it true? Testing claims of any sort is difficult enough. Testing claims about social phenomena is even more difficult. In my *Science as a Process* (1988a), I made a first stab at testing my invisible-hand explanation. I broke down the postulated mechanism into as many localized parts as possible and tested each as separately as possible. For example, might not science work better if it were organized differently? Might it not work better if scientists submerged their individual egos to the general good, as Francis Bacon (1620) suggested? Such contrary-to-fact conditionals are very difficult to evaluate – unless they actually occur – and in the case of the preceding questions they have. The French Academy was formed in 1666 along Baconian lines so that all the credit for the contributions of individual members of the Academy went to the Academy as a whole, rather than to individual members, but by 1699 this noble experiment was abandoned as a failure and replaced by credit for contributions. The Royal Society of London adopted this structure right from the start (Hull 1988a, pp. 322–3).

The one example that I have been able to find in which group effort performed by seemingly anonymous individuals led to major advances in knowledge was the strange case of Nicholas Bourbaki, in which a group of young French mathematicians set themselves the task of reworking the foundations of mathematics. Rather than listing all the individuals who contributed to this project through the years, they invented an imaginary mathematician whom they named Bourbaki. By the time that this project was completed, thirty-four volumes had been published under this pseudonym (Hull 1988a, p. 223).

Even though this example concerns mathematics and not empirical science, it still counts against the mechanism that I am postulating for science because these mathematicians made major contributions to mathematics without (apparently) getting any individual credit. On my view, not all scientists need insist all the time on receiving individual credit for their contributions, but any case of sustained success over long periods by numerous scientists without individual credit does bring my views into question. To be sure, the identities of the changing membership of Bourbaki were well known, and being asked to join this prestigious group was an honor in itself, but I am committed to the view that, upon closer examination, we will discover that these authors did receive individual credit for their contributions (see Weintraub and Mirowski 1994). A graduate student, Michelle Little, is currently working on a case in which one group of radio astronomers held up publishing a major discovery until other groups could catch up to them! If too many cases of this sort can be uncovered, then my hypothesis about the role of invisible hands in science is in serious jeopardy.

5 CONCLUSION

As I understand invisible-hand explanations they do not require that self-interest be the *only* motivation involved, *nor* need all the mechanisms involved in a particular marketplace be totally invisible. As Sen (1983, p. 13) notes, some (partially) planned economies are frequently quite successful. From the beginning conscious effort has played an important role in developing the social structure of science, and such effort has not always been detrimental. For example, when improvements in technology resulted in increased publication delays, date of receipt replaced publication date for awarding credit in science. Is the social structure of science as good as it can get? Probably not, but as the recent history of attempts by the federal government in the United States to set up agencies to police science amply attests, such cures so far have been far worse than the disease (e.g., Kevles 1996).

As science is now practiced, it is a combination of planned and unplanned, conscious and unconscious decisions, but it is the invisible hand that tends to keep scientists on the straight and narrow. The patterns of behavior that I have sketched for scientists are not the result of mysterious coincidences, nor are they optimistic fantasies. To the contrary, the mechanism I postulate is, if anything, overly cynical. What is wrong with invisible-hand explanations? As Ullmann-Margalit (1978, p. 274) notes, nothing, as long as they are both cogent and true. In fact, such explanations are so meritorious that one "can only wish there were more of them."

NOTE

Reprinted with permission from *Philosophy of Science* 64 (1996): S117–S125. © 1997 by the Philosophy of Science Association. All rights reserved.

I would like to thank Toni Carey, Arthur Diamond, Peter Godfrey-Smith, D. Wade Hands, Alistair M. Macleod, Philip Mirowski, J. Tim O'Meara, George Reisch, and Miriam Solomon for reading and commenting on early drafts of this paper.

III

Testing Our Views about Science

7

A Function for Actual Examples in Philosophy of Science

In the early years of science, scientists emphasized, possibly exaggerated, the importance of actual experiments and observations in science in order to free themselves from what they took to be the sterile scholasticism of their predecessors. Their message was, instead of deciding that women have fewer ribs than men on the basis of a priori principles, count. However, thought experiments have also characterized science since its inception. The respective roles usually assigned to actual versus thought experiments is that actual experiments are used to decide truth while thought experiments test our conceptual boundaries, but as Kuhn (1964) has argued in his paper "A Function for Thought Experiments," matters of meaning and truth cannot be so clearly and conclusively distinguished. During the ongoing process of science, they are too closely intertwined. Only in retrospect, when a particular conceptual scheme is thought to be complete, can matters of truth and meaning be separated so neatly.

In this paper, I want to argue for two positions: first, the superiority of actual over imaginary examples with respect to truth in both science and philosophy of science, and second, the superiority of actual over imaginary examples even with respect to meaning in both science and philosophy of science. Except for those who reject the notion of empirical truth *tout court*, no one is likely to object to my claim that only actual counterexamples are relevant to matters of truth in science. No one is likely to reject a scientific theory just because one can imagine evidence counting against it. After all, this is supposedly one mark of a genuine scientific theory. Only actual disconfirmations are relevant to the truth of an empirical claim, and even in such circumstances, it is far from easy to decide when to reject a view in the face of apparent disconfirming evidence. There are always ample opportunities for legitimate finagling. Can speciation occur in the absence of reproductive isolation? It is hard to say, and not all the problems that arise in attempts to answer this question concern evidence.

My claim that data is also relevant to the sorts of observations that philosophers make about science is sure to be more controversial. In fact, a fairly prevalent view of philosophy of science is that it deals only with matters of meaning and never matters of fact. Philosophers provide conceptual analyses – criteria for rationality, good explanations, etc. An analysis of scientific explanation could be retained even if no scientist in the history of science ever lived up to its standards, even if no correlation could be found between the apparent "success" of scientists and the degree to which they approached this ideal. Confronting scientific theories with data is difficult enough. Bringing data to bear on meta-scientific claims of the sort that philosophers make is even more difficult. However, I think that it can be done. Can Mendelian genetics be reduced to molecular biology? It is hard to say, and the problems that arise are not limited just to issues concerning the meaning of "reduction."

My assertion that actual examples are superior to imaginary examples in testing the limits of our concepts is likely to be even more controversial. After all, the grasp of any scientific theory must exceed its reach. However, as far as counterexamples are concerned, nature provides such a rich source of problem cases that rarely will a scientist have to conjure up imaginary counterexamples to add to this store. In fact, scientists are lucky to be able to rework their concepts to accommodate real phenomena without worrying too much about thought experiments. However, scientists do avail themselves of thought experiments, and a few words need to be said about the strengths and weaknesses of such imaginary examples in testing the coherence and completeness of our conceptual systems. Finally, if the philosophical claims made about science are normative rather than empirical, then data are irrelevant. Whatever warrant philosophical claims might have must be elsewhere. In addition to this putative warrant, all that remains are matters of meaning, and here I want to urge once again real over imaginary examples. I begin by discussing truth and meaning in science and only then turn to parallel issues in philosophy of science. Any problems that arise with respect to science are liable to be only magnified when we turn our attention to philosophical, meta-level assertions.

In the ensuing discussion I make reference to some of the papers in

Ruse (1989c) to illustrate my general conclusions. Some of these papers are written by people trained primarily in biology, others by people who are officially philosophers, and others by historians. In my discussion I am interested in activities, not professions. Although I frequently refer to the people practicing these professions, to scientists, historians, and philosophers, my main concern remains the activities regardless of the professional training and affiliation of those engaging in these activities. The justification for this decision is obvious. Although those contributors who are officially scientists differ to some extent in their training and interests from those who are officially philosophers or historians, they are not shy about straying into each other's territories. In fact, the scientists who have contributed to Ruse (1989c) devote more time to discussing philosophical issues (such as the nature of laws) than do the philosophers, and conversely it is the philosophers who pay greater attention to the more scientific issues. And this is as it should be. In his contribution, my good friend of many years, Michael Ruse (1989a), has accurately diagnosed my general views about the relation between history of science, philosophy of science, and science. We are all engaged in the same activity, only with different though complementary training. Increased interactions between scientists, historians, and philosophers are proving productive for all those concerned, and those of us who concentrate on biology have led the way in this symbiotic relationship.

THOUGHT EXPERIMENTS IN SCIENCE

As several authors have emphasized in Ruse (1989c), concepts exist in a matrix of observation statements and results of experiments, on the one hand, and laws and theories, on the other. Even the most abstruse, theoretical claims in science have some observational consequences, and conversely, even the most observational of our concepts in science are to some extent theory-laden. Neither "raw" data nor epistemological "givens" play any role in science. Perhaps we have sense data, but we cannot talk about them. The "holistic" interdependence of observation statements and theories exacts a cost. It makes simple, neat analyses of both the products of scientific investigation as well as this process itself all but impossible. It is also this strong interdependence that makes thought experiments in science productive while the near absence of such interdependence frustrates a comparable use of imagi-

153

nary examples in others areas of investigation. As Kitcher (1989, p. 205) remarks, most of us are "familiar with the dismaying degeneration that characterizes fields in which energy is lavished on counterexamples of no theoretical importance."

Thought experiments to be productive must always occur in a context. They require a "normal situation," the more constrained and detailed the better. Thought experiments help to clarify issues so that future actual experiments can be constructed to distinguish between the most interesting alternatives. However, both actual and thought experiments carry with them a legacy of past experiments and observations. Without such a legacy, thought experiments are of no use in the production of conceptual clarity. Larry Laudan (1989) is certainly right about the effect of the rational weight of the past on science. Furthermore, scientists cannot always anticipate which constraints are going to turn out to be relevant. When scientists finally get around to running what had previously been only a thought experiment, they frequently have to go back and rethink their thought experiments because they had not envisaged certain contingencies.

For example, several of the contributors to Ruse (1989c) discuss the species problem. What are species? What criteria should be used to define the species category? Aborigines have their species concepts, religious fundamentalists have their species concept, alpha taxonomists theirs, the pheneticists theirs, phylogenetic cladists theirs, pattern cladists theirs, and so on. In the absence of a sufficiently specified context, too many answers are possible, and there is no way to decide among them. Just referring to species is not enough. We cannot probe our conceptual limits because there are simply too many of them and most are too amorphous. In such informal, multifarious contexts, imaginary examples do nothing but increase confusion by posing questions that we are not prepared to answer.

For example, people think that they can conceive of a centaur, and in a very superficial way, they can, but too many questions remain unanswered, and no means exist for answering them. How many pairs of lungs does a centaur have, how many hearts? How are the circulatory and pulmonary systems of these creatures connected? What happens to the food that has been digested in the human half of the centaur? Does it empty into the stomach of the horse half? Of course, we are not supposed to ask such questions of mythical creatures. Legends usually tell us enough of what we need to know about centaurs and unicorns for them to play their mythical roles, but sometimes not. When I first came across stories of centaurs as a child, I wondered where baby centaurs come from because apparently there are no lady centaurs.

Only when a context is specified more narrowly can anything very definite be said about species. Because evolutionary theory is currently undergoing a period of fundamental reevaluation, even specifying species as being the things that evolve leaves numerous alternatives open. For the past couple of centuries, the two most common criteria for defining the species category have been descent via sexual reproduction (genealogy) and some sort of morphological gaps between species. More recently, increased emphasis has been placed on ecological considerations. As long as these various criteria always produced the same groupings, biologists were able to postpone opting for one criterion over the others. This situation would seem to be the ideal place to construct thought experiments or, at least, "thought observations." What if organisms existed which appeared in two significantly distinct morphological types but mated freely, always producing offspring of one type or the other? Conversely, what if organisms existed which exhibited no appreciable morphological differences from each other but which never mated successfully even when they had every opportunity to do so?

As it turns out, no such thought experiments are necessary because nature supplies numerous examples of both situations and many more besides. In the absence of such examples, biologists might be forced to invent them, and I am willing to bet that even the most fertile imaginations would not have been able to come up with the bizarre situations that actually occur in nature. These actual examples have additional advantages as well. When questions arise about the particulars of the examples, they can be answered, and these answers introduce issues that would not have been raised by imaginary examples tailored to test only one particular conceptual puzzle. "Why are these two species of fish considered separate species when they mate freely whenever the occasion arises and produce fertile offspring?" Because all their offspring are female. "But why then do not these females mate with the males of the parent species?" They do, but the original hybrid females are so much more vigorous than the organisms belonging to the parent species that they out-compete them in the struggle for life. After several generations, no males remain. Eventually all the hybrid females die as well, and extinction results.

Several of the authors in Ruse (1989c), in discussing the species problem and phylogeny reconstruction, refer to those ubiquitous species A, B, and C and utilize simple "Tinkertoy" diagrams. Does the call for reliance on real examples preclude such devices? Not in the least, because these authors provide real examples as well, examples that can be investigated in greater detail if necessary. The diagrams and simplified examples are meant to be only illustrative, but as Kitcher (1989, p. 206) warns, such simplification has its dangers: "it is easy to draw branching diagrams and to canvass possibilities by appealing to them. But it is always worth asking how we link the organisms that the naturalist observes to the branching diagrams." Such answers necessarily commit the author on a host of issues, issues which in other contexts and at other times might well be brought into question. Real examples allow them to be questioned, but not everything can be questioned at once. For such simplified examples to be of any use, numerous alternatives must be forestalled. The reason that Sober's (1989) Tinkertoy example is so decisive is that he limits himself to a narrow context – cladistic analysis. Cladists do not deny the possibility of anagenic change. They merely decline to divide a gradually evolving lineage into sequential chronospecies. But what if such species are allowed, what if outgroups can be paraphyletic, and so on? Under such circumstances, all bets are off.

In his paper on species concepts, Kitcher (1989, p. 204) presents puzzles for all current attempts to define the species category and finds all proposed definitions seriously deficient. As a result, he urges that we take a "pluralistic view of species, allowing that there are equally legitimate alternative ways of segmenting lineages – and indeed legitimate ways of dividing organisms into species that do not treat species as historical entities at all." The problem with Kitcher's analysis is not that he introduces numerous counterexamples of no theoretical importance. His puzzles are all either real examples or else represent a class of real examples. The problem with Kitcher's discussion is that it lacks a sufficiently focused context. The "current needs of biological research" and the species concept as "naturalists and theoretical biologists alike use it" are not good enough.

Biologists are currently engaged in numerous, semi-independent and partially conflicting research programs, as the papers in Ruse (1989c) amply illustrate. Although members of these research programs refer quite freely to species, it does not follow that they are referring to the same thing or even that they are attempting to converge on a univocal concept. Using the same term to refer to all these various concepts amounts to little more than equivocation. Scientists cannot entertain all alternatives. They must opt. Some biologists have opted for species as lineages. Hence, all sorts of counterexamples cease to be relevant to their species problem. Perhaps these biologists might be led eventually to give up their entire research program, but short of that, certain natural phenomena which would pose puzzles from a different perspective pose no problems for them.

Kitcher is right to expect those engaged in developing a scientifically adequate species concept to provide criteria for making "principled" decisions, but he cannot expect such criteria to apply across all legitimate research programs currently underway in evolutionary biology, let alone all of biology. One price that we pay for scientific advance is that sometimes we must give up some of our preanalytic intuitions. Perhaps many taxospecies are not going to count as evolutionary species, perhaps organisms that reproduce asexually do not form species, perhaps some organisms that reproduce sexually belong to no species whatsoever. These consequences of a particular theoretical perspective are "puzzles" only from other theoretical perspectives or from a more global perspective, usually the hodgepodge of ill-formed, half-articulated, poorly understood conceptions of common sense and ordinary discourse. There is nothing so ill-conceived or poorly understood that someone or other has not been "inclined" to say it.

The main message of the preceding discussion is that real counterexamples pose such serious problems for any species concept that one need not resort to imaginary thought experiments, and that the introduction of such imaginary examples is likely only to confuse the issue. Once a species concept has been devised that can accommodate all the familiar situations that occur in nature, it is time enough to introduce bug-eyed monsters. Furthermore, a species concept will be judged adequate, not from some global perspective but in the context of a particular scientific theory, e.g., some version of evolutionary theory. The puzzles that remain will be as numerous as they are irrelevant. Of course, in a different context, *some* of these puzzles may well turn out to be relevant, and one context can supplant another. If evolutionary theory or some versions of evolutionary theory are abandoned, then which examples are relevant and which irrelevant changes as well. Conversely, empirical discoveries, such as the genetic heterogeneity that characterizes most species, feed back into our concepts. This feedback is only one reason why deciding empirical truth is so difficult.

THOUGHT EXPERIMENTS IN PHILOSOPHY OF SCIENCE

All of the preceding has concerned the role of real versus imaginary examples in science, but the use of imaginary examples is even more prevalent in philosophical contexts, e.g., grue emeralds, clear, tasteless fluids with a chemical make-up XYZ, rabbits named "Snow White," and what have you. Analytic philosophers have shown especially strong partiality to imaginary examples, the sillier the better. One virtue of such examples is that they do not presuppose any technical knowledge. As a result, they are easy to conjure up and can be set out in very few words. Real examples, especially those drawn from science, require extensive knowledge and take up much more space. Goodman (1954, pp. xix–xx) justifies has own reliance on "commonplace and even trivial illustrations rather than more intriguing ones drawn from science" because the examples that "attract the least attention to themselves are the least likely to divert attention from the problem or principles being explained. Once the reader has grasped a point he can make his own consequential applications. Thus although I talk of freezing radiators and the color of marbles, which are seldom discussed in books on chemistry and physics, what I am saying falls squarely within the philosophy of science."

Sometimes philosophers of science have obscured the points they are making by illustrating them with overly technical examples. Simpler examples would do as well. In fact, sometimes the technical discussion serves to obscure the fact that the author has no philosophical point to illustrate. However, regardless of my own prejudices on the matter, one thing is clear: the commonplace, trivial examples that philosophers introduce to illustrate their principles have attracted massive attention to themselves. It is difficult to imagine how more attention could have been paid to a technical example than has been lavished on Goodman's infamous grue emeralds or Putnam's brains in a vat.

Goodman's justification for reliance on commonplace and even trivial examples depends on their being used to *illustrate* preestablished principles. But the examples used by analytic philosophers function as more than illustrations. One sort of exposition that used to be quite common in the philosophical literature was a kind of conceptual striptease. A concept is introduced (e.g., law of nature) and a general analysis is suggested. Successions of counterexamples and modifications are then introduced until either an adequate analysis is reached or else the original concept degenerates into a conceptual morass. In a single paper, the first alternative tends to prevail. However, as successive authors address the issue, the second alternative begins to predominate. One might think that a significant difference exists between the claim that all the coins in my pocket are American and Newton's law of universal gravitation, but that belief, so it seems, is sorely mistaken. All is one – or nothing, which amounts to the same thing.

Some of the examples that philosophers introduce into their conceptual deconstructions may serve as nothing more than illustrations, but others also function as justifications. For example, a common argument in the philosophical literature is that not all effects are functions because the sounds that hearts make as they pump blood are effects and not functions. This example and others like it not only illustrate a point but also serve to justify it. I find the basis for such justifications problematic. Because the examples are commonplace, the only foundation that seems warranted is that provided by common sense and ordinary experience, and in my book both are extremely untrustworthy. I much prefer grounding philosophical analysis of science in science, not common sense, and such grounding requires technical examples.

In science, examples function as more than illustrations; they also serve as evidence, and one thing can be said for certain about silly, science fiction examples: they cannot serve as evidence. Of course, if nothing like evidence can be brought to bear on philosophical claims, this characteristic of imaginary examples hardly counts against them. Years ago, when I was even more naive than I am now, I thought that claims about the reducibility of Mendelian genetics to molecular biology had something to do with what was going on in biology at the time (e.g., the genetic code, molecular mechanisms, reaction norms) and argued that *in point of fact* a Nagel-type reduction from Mendelian genetics to molecular biology was impossible (Hull 1972, 1973). Others responded by modifying and expanding upon Nagel-type reductions, attempting to show that even in the face of the empirical facts of the matter, reduction was possible. Such modifications were perfectly legitimate. They are exactly the sorts of modifications that data are supposed to elicit. If biologists can legitimately modify their gene concepts as they find out more about the genetic material, then philosophers can certainly modify their concept of reduction in the face of increased knowledge about putative cases of reduction. In Ruse (1989c), Rosenberg (1989, p. 248) carries this process even further, showing the implications of more recent advances in molecular biology for any philosophical notion of reduction. As Rosenberg sees it, "smooth" reduction will not go through in biology because selection for function is blind to structure: "*it cannot discriminate between differing structures with identical effects.*"

In science data leads scientists to modify their theories, but if we accept the thesis that even the most observational of statements are to some extent theory-laden, then no such thing as "raw data" exists. The theories that we entertain influence how we describe natural phenomena. Parallel observations should hold for philosophical theories about science. As Lakatos (1970) has argued persuasively, no one can write history as it was without any philosophical presuppositions. Inductivism with respect to the course of science is even less justified than inductivism with respect to the genesis of Kepler's laws. If scientists cannot merely "sum up observations," then historians should be unable to write histories of science that record the facts and nothing but the facts.

Several of the papers in Ruse (1989c) address these parallel problems in especially pointed ways. For example, Cracraft (1989, p. 39) argues for a particular definition of species which combines ontological and epistemological considerations. Strange as it might sound when put so bluntly, species are observable individuals. They are observable in the sense that they are the "smallest clusters of individual organisms sharing diagnostic character variation." The characters in question are properties of organisms. Of course, a human being cannot literally observe character variation in species containing millions of organisms spread over thousands of square miles, let alone millions of years. A systematist observes these characters in a small percentage of organisms and *infers* patterns of variation. Cracraft also insists that species exhibit reproductive cohesion. They must be self-perpetuating. Cracraft takes this characteristic of species to be so basic that he does not bother even to argue it. One consequence of this assumption is that males, females, juvenile stages and various morphs are included in the same species even if they lack the requisite diagnostic characters.

Although Cracraft acknowledges that the contrast between species as observables and species as theoretical entities is far from sharp, he opts for species as belonging nearer the observable end of the continuum. At the very least, any appropriate definition of the species category should not presuppose anything about the evolutionary process.

Reproduction and ontogeny are acceptable; evolution is not. Perhaps a species category can be formulated that is not laden with evolutionary theory, but it is nonetheless as theory-laden as a term can get. The main motivation for allowing certain theories to affect one's definitions but not others seems to be epistemological. For example, rarely can human beings hope to observe species evolving, speciating, and going extinct. For extant species, we can observe mating, replication, and ontogenetic development at the organismic level. We *can* observe such events at the organismic level, but in point of fact we rarely *do* – even for extant species. As Cracraft notes, most systematists study character distributions and that is all. There are simply too many species and too few systematists to do more. Species are going extinct faster than we can identify and classify them. As far as extinct forms are concerned, both organismic and species-level characteristics are in the same boat. They cannot be observed.

This much being said, why the heavy emphasis on the part of scientists such as Cracraft on observability? I suspect most philosophers would argue that theoretical definitions are being confused with criteria of application, sometimes termed "operational definitions." Both are necessary, but criteria for application should be appended to definitions, not included as part of them. On the main, philosophers have not concerned themselves with the complex set of problems surrounding the operationalization of theoretical definitions. For the most part, we are content to observe that theoretical terms cannot be defined operationally and let it go at that. But if we are genuinely interested in science, we are mistaken to pay so little attention to matters that consume so much of scientists' time and which they take to be so important. Theoretical definitions combined with lists of techniques for operationalizing them are not good enough. Scientists feel compelled to include operational criteria in their definitions. If data is relevant to the observations that philosophers make about science, then the insistence of so many scientists on retaining certain practices that philosophers take to be so clearly mistaken cannot be dismissed lightly. If scientists were unaware of what philosophers have to say on the subject, we might conclude such mistakes are merely a function of scientists' being unaware of the error of their ways. But as Cracraft clearly shows in his paper, he has read the relevant philosophical literature and still he persists. And Cracraft is not alone in the tenacity with which he resists positions widely accepted in certain philosophical quarters.

Thus, those of us who think that data about science is relevant to philosophical views about science are caught in a bind. We cannot very well use the concordance between scientific practice and our general analyses as support for our views while insisting that departures do not count against us. One plausible solution is that scientists who practice what we preach are, on the average, more successful than those who do not. Problems about what counts as success in science to one side, thus far no one has undertaken such an extensive survey. However, if such a massive effort is to be undertaken, those of us who set out general analyses had better be prepared to have the results make a difference. Perhaps we do not have to take even successful scientific practice at face value, but it must count for something. We cannot dismiss all potential disconfirming evidence in advance.[1]

As troublesome as the preceding problem may appear for meta-science, it is only a repetition of the interplay between theory and observation in science. Inductivist scientists want to keep observation reports totally free of the theories that these reports are going to be used to test; otherwise they fear that they will be caught up in circular reasoning. Philosophers are quick to point out that the sort of "reciprocal illumination" that results from the interplay between observation and theory construction need not lead to vicious circularity, but we do not seem to appreciate how disorienting this procedure can get in particular cases. For example, if a systematist distinguishes characters in such a way that they form transformation series of primitive and derived characters, then he or she will produce nested sets of characters. The fact that these characters do nest so neatly thereby indicates that they were individuated correctly to begin with. Any characters that do not fit are not really characters. If such reasoning is legitimate in science, then it should be equally legitimate when applied to science itself, uncertainty about particular cases notwithstanding.

One reason why the inferential interplay between theory and observation is so unsettling is the fear that the theoretical component of observation statements will be so strong that we will never be able to discover when our theoretical views are mistaken. Although we can see in retrospect that in some cases our allegiance to particular theories has frustrated attempts to falsify these theories, in other cases scientists strongly committed to particular views have been forced to reject or modify them in spite of their allegiance. Observation statements are theory-laden but not so laden that they cannot refute the very theories which color them. Parallel observations hold for incommensurability.

In the context of a particular semantic theory, observation statements derived from different scientific theories cannot contradict each other because certain key terms in them do not mean precisely the same thing. However, when this semantic thesis is operationalized to refer to actual scientists, it turns out to be false. Adherents of two different theories should not be able to agree on observations that would refute one theory and support the other, but they can and do. As different as William Thomson's thermodynamic theory was from Darwin's theory of evolution, the implications of these theories for the age of the earth were so different that no one could avoid the conclusion that at least one of these theories had to be seriously in error.

As is the case with so many historians, Hodge (1989) is unhappy with the use that philosophers have made of his favorite scientist. On the basis of a lifetime of philosophically informed historical research, Hodge has worked out the structure and strategy of Darwin's argumentation, not just in the *Origin of Species* but in related works as well. Philosopher after philosopher has attempted to support his or her philosophical thesis by appealing to Darwin. Hodge responds that, damn it, you have gotten Darwin wrong. One possible response to Hodge's complaint is that no description of the course of science can be totally free of general philosophical views about science. The reason that Hodge thinks that these philosophers are mistaken about the structure and strategy exhibited in Darwin's work is that he has reconstructed it on the basis of inappropriate philosophical or metascientific presuppositions.

This response is surely more appropriate for some historians than for others. It might be the case that the differences in general philosophical outlook between Hodge and those who want to use Darwin to illustrate (not to mention support) their general philosophical views are so subtle that the facts of the matter cannot possibly resolve these differences. Conversely the facts of the matter might be so apparent that they can be decisive. Although our meta-theories color our descriptions no less than do the more pedestrian theories constructed by scientists, sometimes even these highly theory-laden descriptions can refute the theories that laden them. For example, even if Mendelian genetics and molecular biology are reconstructed to fulfill the needs of Nagel-type reductions, the necessary correlations between Mendelian and molecular natural kinds are not forthcoming. Perhaps such a state of affairs is impossible, but it does occur. Hodge can be read as making just such a claim for Darwin and the philosophical theses that Darwin's work is supposed to exemplify. For any philosopher who claims that at least sometimes data about science can be brought to bear on at least some of the philosophical theses which philosophers of science set out about science, complaints by historians such as Hodge cannot be dismissed lightly.

CONCLUSION

In this paper I have questioned the independence of two traditional philosophical dichotomies: between truth and meaning and between observation statements and theories. Issues of truth can cause us to change what we mean by what we say just as surely as the meanings that we assign to the terms we use influence what we take to be the case. One might be tempted to dismiss such holistic interdependence as being characteristic only of the ongoing process of knowledge acquisition. As in the case of government in the perfect Marxist state, this interdependence will wither away once total knowledge is reached. Because I think that we will never reach this ideal epistemic state, I am modest enough in my ambitions to limit my observations to the only state of affairs that will ever obtain.

Even though I think that some philosophical theses can be interpreted in such a way that data can be brought to bear on them, others are liable to be more than a little recalcitrant. For example, although I think that Rosenberg (1989) shows the clear implications that the current state of genetics and molecular biology have for Nagel-type reductions, I doubt that anything that might count as data or evidence could possibly influence the debate over instrumentalism and realism. These positions have become way too sophisticated. No matter what course the history of science takes, all sides will be able to claim victory. In any case, one promising avenue of research for those of us who are interested in science is to look for general views about science that can be tested and to test them. For example, how important are various sorts of influences on the decisions that scientists make? Do such "external" factors as the social structure of scientific communities or society at large influence the course of science? Grene (1989) fears that such concerns might lead to the abandonment of an inadequate philosophy for none at all. I think not, but we will have to wait and see.

Kitcher (1989) expects scientists to present principled ways for drawing the distinctions that they take to be central to their areas of investigation. Although scientists might have to change these principles as time marches on, ad hoc, case-by-case decisions are not good enough. I agree, but these principles are not going to apply across all contexts, even across all legitimate contexts. Moving to the meta-level, philosophers should also be expected to present principled ways to draw their distinctions as well. For example, Kitcher is not a promiscuous pluralist. He argues that certain species concepts are illegitimate, in particular the species concepts urged by creationists and pheneticists. What is needed is a principled way to distinguish between legitimate and illegitimate scientific concepts. I suspect that whatever principles Kitcher devises, they will not apply across all philosophical contexts. Just as scientists must opt, so too must those of us who study science. One reason for picking one avenue of investigation over another is recent success. How much progress have phenomenologists made in their goal of reconstructing science on an adequate phenomenological foundation? As far as I can tell, not much. Have more analytically-inclined philosophers made much progress in their quite different goals? I am certainly biased in this respect, but as far as I can tell, quite a bit, and those of us who have concentrated on biology have done more than our share in contributing to this success.

NOTES

Reprinted from *What the Philosophy of Biology Is: Essays Dedicated to David Hull*, edited by Michael Ruse (Dordrecht: Kluwer Academic Publishers, 1989), pp. 309–21, with kind permission from Kluwer Academic Publishers.
1. The uneasiness that philosophers are sure to exhibit toward the possible implications of statistical studies of science for their philosophical analyses has been expressed recently by psychotherapists when the government proposed to study the efficacy of various sorts of therapies. If government money is used to pay for such therapy, then some evidence must be provided that it does some good on some definition of "good." Psychotherapists responded that the results of such studies, no matter how they are conducted, cannot possibly indicate anything about the efficacy of psychotherapy. I suspect that this reflex action is more a function of the insecurity that psychotherapists feel about the efficacy of their therapies than problems in research design.

8

The Evolution of Conceptual Systems in Science

Prior to Darwin linguists commonly treated languages as evolving systems. In fact, Darwin used this analogy to lend credence to his theory of the evolution of biological species, but since 1859 much more sustained attention has been devoted to how species evolve than to how languages evolve. The rise of sociobiology almost two decades ago directed attention once again to evolutionary explanations of behavior, especially human behavior. One of the most important characteristics of the human species is the degree to which we can communicate with each other. Our ability to learn about the world in which we live is our most important evolutionary advantage, and this ability could not be realized to the degree it has been without the sort of social learning that communication permits. Nowhere have the techniques of social learning been more highly developed and explicitly codified than in science. For this reason, the topic of this paper is the evolution of conceptual systems within science.

Numerous papers and books have been written arguing that conceptual change can profitably be viewed as an evolutionary process in which selection of sorts plays a central role. If certain ways of thinking are sufficiently adaptive from a strictly biological point of view for long enough periods of time, these conceptual predispositions can become programmed into the genes responsible for the formation of our nervous systems. For example, people seem to be born strongly predisposed to essentialism, i.e., to view the world as being divisible into kinds that can be characterized by universally covarying characteristics. Man is a rational animal, and all rational animals count as being human. Even so, we cannot help but notice the variation that surrounds us – it is too apparent to ignore totally – but we automatically transform this variation into deviation from a norm. For example, any human being who is incapable of rationality is not normal, not fully human, and the behavior of other organisms, no matter how rational it might seem, is dismissed out of hand. Imposing overly discrete boundaries on a variable world seems to have been a sufficiently adaptive response often enough, over a long enough period of time, to have been made part of our neural make-up.

However, many of our beliefs are very transitory. Certain beliefs may be advantageous for a while, then disadvantageous. This sort of information is programmed into our brains, not our genomes. Such beliefs are transmitted socially, not genetically. Gene-based biological evolution is driven by a selection process. How about individual learning? Social learning? If gene-based biological evolution is taken as the paradigm selection process, then meme-based conceptual evolution must be viewed as merely an analogous process. The characteristics of biological evolution are essential to selection processes, and any variations from this norm are deviations. For example, according to the theory of clonal selection, the reaction of the immune system to antigens is a selection process, but since clonal selection deviates from gene-based biological evolution in several important respects, it is at best an abnormal example of selection, at worst merely an analogy (for a contrary view, see Darden and Cain 1989).

Comparable objections have been raised to conceptual selection. Since any departures from gene-based biological evolution are deviations from the norm, and conceptual change does differ at several points from biological evolution, it too must be judged to be at best a deviant form of selection. Even though viewing the world in terms of paradigm cases and monsters may well be programmed into our genomes, the reaction norms for this predisposition are wide enough for it to be overcome, but like all bad habits, it reasserts itself when we are not paying sufficient attention. The happenstance that selection in gene-based biological evolution was the first selection process to be studied in any detail does not make it the standard against which all other forms of selection must be measured. A much better way of treating selection is to look at all the natural phenomena that appear, on the surface, to be instances of selection to see what characteristics they exhibit, and then decide which of these characteristics are important, possibly essential, and which are not. Among these phenomena are biological evolution, the reaction of immune systems to antigens, neuronal selection, connectionism in cognitive psychology, operant conditioning, and conceptual change in general (for a description of neuronal selection, see Plotkin 1991).

After such a sampling, a more broadly based account of selection can be given. On this account, replication turns out to be central to selection, but the entities being replicated need not be made of DNA or RNA. Generations are also necessary, but the generations need not be organismal. Numerous cycles of selection can occur during the space of a single organismal generation. In addition, certain entities must interact with their environments so that replication is differential, but these entities need not be organisms. For example, the interactions that influence biological evolution occur at a wide variety of levels, from molecules and cells to colonies and demes. Selection as a general process results from an interplay between replication and environmental interaction (Hull 1988a). The question then becomes whether or not conceptual change can be interpreted to fulfill these more general conditions.

Numerous authors have set out theories of how conceptual change can be construed as being a selection process. The conceptual problems involved in such an enterprise have been discussed far past the point of diminishing returns. Critics are sure that they have conclusively refuted this research program, as conclusively as every other research program in human thought has been refuted, including those that eventually came to prevail. I take such refutations none too seriously. However, those of us who think that conceptual evolution can be treated profitably as a selection process are at a juncture. We need to put up or shut up (Heyes 1988, p. 194), to put our money where our mouths are (Ruse 1989b, p. 221). For those of us who are interested specifically in the conceptual development of science, we "must look in depth at episodes in the history of science, or ongoing disputes and movements in contemporary science, and see if fresh and valuable perspectives can be gained by employing the evolutionary model. . . . At this point therefore all I can urge is that the enthusiasts stop talking and get to work" (Ruse 1989b, p. 221).

I could not agree more. We have argued long enough about the *possibility* of treating conceptual evolution as a selection process. It is about time we paid greater attention to delivering on our promissory notes. We need to develop models and test them to see which of the possibilities turn out to be actualized. Many hypotheses about science seem plausible enough. Others seem very counterintuitive. We need to know which of these various hypotheses are true. Examinations of particular episodes in science help, but we also need large-scale, summary data of the sort that have been collected to test comparable hypotheses in evolutionary biology. Although we need to test our hypotheses about science, we also should not set up unrealistic standards. Much of the evidence that purported to test early formulations of evolutionary biology turns out to be erroneous or not all that relevant (Dobzhansky 1937). Certain parts of evolutionary theory remain largely untested, the evidence for other parts is extremely equivocal, while other parts have ample evidential backing (Endler 1986). We should not evaluate emerging research programs on standards more stringent than well-established research programs can meet.

Conceptual evolution as such is a very large topic. In my own work, I have limited myself just to conceptual evolution in science, a sufficiently daunting topic on its own. Conceptual evolution in science is in no way a paradigm of conceptual evolution in general, but it provides an especially good test case for theories about conceptual change because so much of it is conducted in the open and the recorded data are so extensive and readily available. In this paper I investigate two mathematical models that have been devised to handle conceptual change in science and the extensions of several others to science, even though they were not specifically devised with science and scientists in mind. I then turn to the issue of actual testing.

MODELS OF CONCEPTUAL SELECTION IN SCIENCE

Selection processes require that replication be differential. Not every replicate that is produced can survive to reproduce. Scarcity of some sort is required. In conceptual evolution, words are cheap. Even so, much of the research conducted by scientists never sees print. Of the manuscripts that do make it through the refereeing process, only a very few seem to have any impact on the course of science, and actual usage is what really counts. Explicit citations form one reasonably good measure of use in science, but citations also play several other roles as well. One of them is to indicate agreement and disagreement. Another is to leave a paper trail for assigning credit and blame. But one of the most important functions of citations is to garner support. In the most highly developed areas of science, a significant majority of citations are positive in the sense that they specify the previous work on which the current research is based. The more that researchers can show that their work is grounded in the good solid work that has preceded it, the more seriously others will take their results. The price that they pay for such support is loss of credit for innovation and originality. Scientists desire credit and need support, but with respect to particular decisions to cite or not to cite, they cannot have both.

The low-risk, low-reward strategy is to make one's own contributions look as firmly grounded in the preceding work as possible. The high-risk, high-reward strategy is to include as few positive references as possible in order to make one's own findings appear as original as possible. Of course, this strategy also increases the likelihood that one's work will be ignored, both because it will seem to lack adequate grounding and because the authors who have been ignored are likely to return the favor. This story is complicated by the fact that not all scientists are equal. The support of a major figure in the field is worth much more than the support of a graduate student or postdoc. These younger workers are also very vulnerable because they are not in a position to insist that they receive the recognition that they think they deserve. Hence, one would expect authorities to be cited much more often for their support than their less visible colleagues, the intrinsic merit of their work to one side (Hull 1988a).

The preceding observations about the interconnections between use, citation, credit, support and authority may sound plausible enough, but are they true? The first prerequisite for answering such questions is the formulation of explicit models. Using the principles of rational decision theory, Kitcher (1992) has produced a model designed to test under what circumstances individual scientists are likely to show deference to authority and what effects this deference has on science. He also attempts to estimate such things as the conditions under which all scientists working in a particular area will ignore a challenging finding, universally embrace it, or selectively test it. On his model, the different responses to the claims about cold fusion made by Stanley Pons and Martin Fleischmann by electrochemists versus physicists are explicable in terms of differences in the cognitive authority of these authors in these two communities.

The mathematical models that Kitcher presents are extremely useful because they indicate how certain sorts of behavior are interrelated. Of course, his models are overly simple and artificial. Such models always are. In anticipation of such complaints, Kitcher compares his models to those produced in the early years of population genetics by R. A. Fisher, J. B. S. Haldane, and Sewall Wright. They too were artificial and overly simple, but they led to the development of evolutionary biology as we know it today. The hope is that comparable models will lead to a more highly developed science of science (for further defense of using models to study complex phenomena, see Richerson and Boyd 1987).

One of the most important characteristics of the social structure of science is that it is so organized that attempts to increase one's professional success tend to promote the acquisition of reliable knowledge. When scientists produce results that other scientists find useful, by and large they use them. If these results turn out to be faulty, the careers of the scientists who use them are damaged. Months, perhaps years of labor are wasted. Scientists who find themselves in such a predicament follow the paper trails provided by citations in order to discover the culprit. The initial punishment in such cases is the publication of a paper in which these early results are shown to be mistaken. The more general punishment is a decrease in the use by other scientists of the work produced by such unreliable scientists. Since use is the chief coinage in science, the refusal to incorporate the work of a particular scientist in one's own work is the worst punishment possible (Hull 1988a).

The desire for credit *can* foster truth acquisition, but it need not. At times, it can interfere. Goldman and Shaked (1992) compare the effects of credit-seeking and truth-seeking motivations in the framework provided by Bayesian decision theory. In their terminology, when a scientist revises his or her subjective probabilities in light of the work of another scientist, he or she credits this second scientist with a contribution. Goldman and Shaked conclude that credit motivation does foster truth acquisition in most circumstances but that credit motivation alone does not serve this epistemic goal quite as well as truth motivation alone. The most important cases, of course, are those in which scientists are motivated by both considerations. They want to solve particular problems, *and* they want other scientists to acknowledge their contributions by incorporating them into their own work. As in the case of Kitcher (1992), the strength of the models produced by Goldman and Shaked is that they indicate which circumstances are relevant and how they are likely to influence the outcomes.

The preceding models were designed specifically to deal with conceptual change in science, but models devised to deal with people in general frequently apply to scientists as well. For example, Boyd and Richerson (1989) investigate the circumstances in which cooperation can occur among genetically unrelated individuals – the typical case in science. If the only response to uncooperative behavior is withholding future cooperation, then Boyd and Richerson show that indirect reciprocity can become important only in relatively small, closely-knit, long-lasting groups. However, if the possible responses to uncooperative behavior are expanded to include selective punishment, then Boyd and Richerson show how cooperation can evolve in much larger groups. As they point out, punishment costs not only the person being punished but also the person administering the punishment. If the only good that results from such punishment is for the group at large, then selection should favor individuals who cooperate but do not punish. Only if punishments are very severe and individuals are punished for not punishing can moralistic societies of the sort with which we are familiar develop.

Although Boyd and Richerson intend their argument to apply to gene-based biological selection, I think it can be expanded to meme-based cultural selection as well. Self-policing professions are infamous for not policing themselves and for exactly the reasons that Boyd and Richerson indicate. In one recent study conducted by researchers from the Harvard Medical School of 51 hospitals in the state of New York, doctors' negligence was found to have caused 27,000 injuries and 7,000 deaths in a single year. Yet these cases led to reprimands in less than one percent of the cases. Medical doctors rarely punish their fellow doctors no matter how serious and frequent the malpractice might be because the only benefit that results from one doctor informing on another is the undifferentiated good of the profession. Nor are medical doctors punished for failing to turn in their fellow doctors for malpractice. In fact, they are frequently punished for such acts. Thus, the failure of doctors to police themselves for the good of the profession should occasion no surprise. Similar observations hold for lawyers, law enforcement officers, and college professors in their roles as teachers.

However, with respect to the publication of faulty results in science, scientists police themselves very well indeed. Any scientist who finds that he or she gets very different results from those in the literature publishes his or her findings. If these new results hold up to scrutiny, they are used by later workers and the erroneous results are gradually displaced in the literature. Of course, scientists do not police themselves perfectly. Errors can remain undiscovered in the literature for years, especially in scientific backwaters, but scientists police themselves much more scrupulously in this respect than do the members of any other self-policing profession. Granted, scientists do not have to police themselves all that well to police themselves better than other self-policing professions, but the differences are significant. Why do scientists police themselves so scrupulously when it comes to errors in the literature? Because it is in their own best self-interest to do so.

It should also be noted that scientists are not nearly so good at policing themselves with respect to unacknowledged use and outright fraud. The use of someone else's results without giving ample credit is certainly decried in science, but it is not punished all that severely. Erroneous results harm everyone who uses them. They threaten everyone working in a particular area of science. Because of the use that scientists make of each other's results, failing to report an error can harm one's own research in unanticipated ways. Even though you are aware that a particular result is spurious, you might well use the work of other scientists who are not aware of this mistake. Hence, it behooves scientists, for their own good, to point out any errors that they find. Passing off the work of someone else as your own hurts only the person whose work has been stolen. Only if such failure to give credit were to become widespread would science itself be threatened.

Fraud presents quite a different set of problems. Science is well-organized to detect errors, but not to distinguish fraud from errors because this difference depends on the distinction between intentional and unintentional actions. A common element in social mores as well as explicitly codified laws is that intentional acts deserve more extreme rewards and punishments than those that are unintentional. Accidentally killing another person is not as blameworthy as doing so intentionally, although the person is just as dead in both instances. When carried over to science, this conviction implies that fraud is a more serious infraction of scientific ethics than unintentional error, even though the effects of these errors are indistinguishable. The publication of a mistaken reading has the same effects whether the error arose intentionally or unintentionally. However, none of the practices that have grown up in science since its inception are well-calculated to make this distinction. As a result, cases of alleged fraud tend to be handled quite badly. If anything, the institutional structure of science is designed to deflect charges of fraud, especially when a highly respected member of the scientific community is involved. When the issue arises, the temptation is to conclude any faulty reports were the result of honest mistakes or possibly sloppy work but not fraud. Whistle blowers in science are treated almost as callously as elsewhere. The recent case of David Baltimore, Thereza Imanishi-Kari, and Margot O'Toole is a good case in point (Hamilton 1991a).

GATHERING DATA TO TEST MODELS OF SCIENCE

Some of the preceding observations might sound plausible enough, while others may seem questionable. The important question is, are they true? In an evaluation of citation patterns dealing with the work of Merton on deviance and anomie, Cole (1975, p. 209) is surprised to discover how little effort had been expended to test Merton's theory. For a period of twenty-two years between 1950 and 1973, 123 articles in four leading sociological journals dealt with Merton's work. Of these, only nine articles reported research designed to test Merton's views, while another 21 showed how data already at hand could be interpreted in the light of Merton's theory. An even more recent example shows how long a belief can be transmitted in science without anyone bothering to test it. For generations the similarity in appearance of viceroy and monarch butterflies has always been presented as a paradigm case of Batesian mimicry. Birds tend not to eat viceroys because they mimic foul-tasting monarchs. As it turns out, viceroys taste just as bad as monarchs (Ritland and Brower 1991).

If we are to ever have a science of science, we need both general theories and more limited models, but we also need data of the sort that we can use to test these models and, hence indirectly, the theories that they illustrate. Since the slippage between theory, models and data is extensive in all areas of science, it should be no less significant in the study of science. Rarely do scientists have anything like direct access to the variables that they want to measure. Instead they have to make do with indirect estimates. Temperature is an important physical variable, and physicists have devised numerous ways to measure it, none of them perfect in all circumstances. Early measures were especially crude, but without these crude measures, the more precise instruments available today would never have been developed. Evolutionary biologists have been presented with comparable problems in estimating fitness, a notion that is equally central to their theories. The literature on IQ tests as measures of intelligence is even larger and more contentious.

The operationalizing of concepts necessary for testing is inherently a matter of compromise. We think that we understand what it is that we want to measure. We also are aware that the techniques that we have developed thus far are not good enough, but the hope is that we will devise better techniques as we proceed. We also turn to where data are most readily available even if these data do not bear on the issues that interest us most. One feature of this process that we tend not to notice is that the results of our admittedly crude measurements frequently feed back into our understanding so that we change our concepts. We may think that we know what we mean before we attempt to test our claims, but the requirements of testing invariably force us to clarify our ideas considerably. Those of us studying science might think that we have a clear understanding of the hypotheses that we have formulated about how science works, but unless we are very special indeed, our conceptions are extremely hazy. Attempting to operationalize these concepts so that we can collect data about them is one of the best ways to clarify our understanding.

Extensive data are available about science. Much of it is buried in the archives of journals and funding agencies. Some, like that in science citation indices, is readily available. However, most of the quantitative studies of science that have been conducted thus far were not devised to test hypotheses about science as a selection process. If they are to be used, they will have to be coopted and reworked for our purposes. Data to support the claim that science is as high-resolution an activity as professional music, sports and ballet is overwhelming. When most of us hear the term "scientist," we think of research scientists churning out paper after paper, but most people who are officially labeled scientists publish very little in their lifetimes. For example, half of the physiologists studied for a three-year period published fewer than four papers during that period (Meltzer 1956). Of 238 chemists receiving their doctorates between 1955 and 1961, 7.5 percent published nothing in the first decade after receiving their degree and 11 percent published only one article. In any given year, 80 percent of chemists publish nothing whatsoever (Reskin 1977). During a two-year period, almost a third of the physicists teaching in American universities and colleges published nothing (Ladd and Lipset 1977). Social scientists are even less productive than physiologists, chemists and physicists (Babchuck and Bates 1962, Yoels 1973). For example, only 10 percent of psychologists publish at least one article a year (Garvey 1979).

Obviously, the dictum that scientists must publish or perish does not apply to a great number of scientists, especially those working in industry, defense, and government. Perhaps citations are not a very good measure of the contributions that scientists make to science, but very few innovations that fail to make it into print are likely to have much of an impact. When we turn to the effects of those papers that are published, the disparity becomes even greater. Two percent of federally employed scientists working in oceanography account for 65 percent of citations in the field (Menard 1971). Of the 10 percent of psychologists who do publish, only 10 percent receive at least one citation a year. Only 10 percent publish at least one paper a year, and only 10 percent of these are cited! An intense investigation of high-energy physicists revealed that only 43 percent contributed anything to either the literature or to administration (Blau 1978). To make matters even more extreme, the same small group of scientists produce not only the few highly cited papers but nearly all the uncited papers as well (Cole and Cole 1973).

Although MacRoberts and MacRoberts (1984) sound a note of caution about how some of these data were collected, the conclusion is difficult to escape that research science is a highly elitist activity. Most scientists publish very little during the course of their careers, and of those who do, most have no discernible influence on the course of science. More recent figures bear out these early findings. For example, *Science* magazine commissioned David Pendlebury of the Institute for Scientific Information to search the data base of this institute to get more general figures on citation patterns. ISI includes information in their data base on only the top 6 percent of scientific journals being published. Pendlebury discovered that 55 percent of the papers published in these most highly cited journals received no citations (including self-citation) in the five years after publication (Hamilton 1990, 1991b). This finding is especially damning because rates of initial citation drop off markedly after the first five years. A few of the papers cited extensively during the first five years after publication continue to be cited after five years, but only a handful of papers not cited during the first five years are ever resurrected by later workers.[1]

Although Cole and Cole (1972, p. 370) note that a "significant minority of cited work is being produced by non-elite physicists," they also suggest that perhaps the "number of scientists could be reduced without affecting the rate of advance" (Cole and Cole 1972, p. 372). Since multiple discoveries are so common in science, the 15 to 20 percent of the contributions being made by non-elite scientists is likely to be duplicated by the elite scientists. Even when it comes to teaching, nearly all the elite scientists were themselves taught by elite scientists. The only real problem that Cole and Cole foresee is how to identify which students are likely to join the ranks of elite scientists prior to society's investing so much of its resources into their training. The best sign of future productivity is how soon a young research worker starts publishing. The earlier scientists begin publishing, the more likely they are to continue publishing throughout their careers. Although there are a few late bloomers in science, there are not many (Cole 1991).

One can imagine the uproar that Cole and Cole's paper elicited, especially since it appeared in a prestigious, widely-read journal. Of course, Cole and Cole were only floating a theoretical possibility. The likelihood that any concerted campaign would be launched to identify those budding young scientists who are unlikely to contribute to the course of science and to discourage them from pursuing a life of science was and remains quite low. However, unbeknownst to the Cole brothers, nature was already in the midst of performing an experiment to test their hypothesis. The exponential growth of scientists in the developed countries that began in the 19th century tailed off drastically in the later 1960s and actually dipped in the 1970s. For example, the number of new assistant professors of physics hired peaked in 1966 at about 400 but dropped to less than 200 in 1972 and stayed there for the next three years. If (and it is a big "if") a disproportionate number of less committed and intellectually duller physicists dropped out during this period, one would expect no comparable drop in productivity. Such turns out not to be the case. The number of papers by young physicists receiving one or more citations in the two leading physics journals dropped markedly in 1970 and did not recover during the ensuing six years. Scientific productivity seems to be correlated linearly with the number of scientists working in an area. Thus, Cole (1991) was forced to conclude that the earlier recommendations by Cole and Cole (1972) were faulty.

What implications do the preceding data have for science as a selection process? The first is obvious: selection pressures in science are extremely intense. In biological evolution, at least, the more intense selection is, the more rapid evolutionary change is. Of course, selection can become so intense that a lineage goes extinct. What are optimal levels for selection in science if what we want is a maximum production of high-quality work? We do not know. One thing that we do know about selection processes is that they are inherently inefficient.

Because contingency plays such a large role both in the generation of novelty and in its selection, large numbers are required. Small numbers increase the likelihood that new species will evolve, but they increase the likelihood even more that these incipient species will go extinct. As a result, attempts to increase the efficiency of selection processes are likely to be quite dangerous. Calls to make science more efficient might sound plausible enough, while comparable calls to make the production of great works of art more efficient might not sound quite so sensible. Science is inherently a creative process, and creative processes are inherently inefficient. One explanation for this inefficiency is the large role that selection plays in any creative process.

THE DEMIC STRUCTURE OF SCIENCE

According to one prevalent view of biological evolution, speciation occurs most rapidly when species are subdivided into numerous, partially isolated demes. A recurrent criticism of science is that scientists form old boy clubs. Well-placed scientists get more money, better facilities, more good students, more awards, etc., than the quality of their work alone would justify, and they pass around these advantages to each other. The fear is that white, middle-aged men predominate in science for the same nonegalitarian reasons that they possess most of the power and benefits in all developed countries. The data on this score are decidedly mixed. Cole and Cole (1973) come down strongly on the side of "universalism" in science. The chief reason why more of the rewards in science go to a small percentage of scientists is that they are doing the vast majority of good work. Although Cole (1991) still thinks that the quality of scientists' work is the major determinate factor in the use that other scientists make of it, he now sees more significant roles for more "particularistic" factors in such matters, in particular variations in the social structure of science (see Hull 1990).

If science is a selection process, the existence of small groups of scientists who pool their conceptual resources to solve particular sets of problems should be highly beneficial (Giere 1988). Such research groups should be the primary sites of conceptual change. Old boy clubs are harmful only if membership is determined by factors unrelated to science. Although commentators on science, not to mention scientists themselves, frequently decry "factionalism" in science, it is a necessary consequence of what I have termed the "demic structure of science." Scientists are intent on increasing their own conceptual inclusive fitness. They get credit not only for their own contributions but also to some extent for those of their co-workers. Coauthoring papers and mutual citation increase scientists' conceptual inclusive fitness. The goal of increasing one's conceptual inclusive fitness is accomplished best, not by working in isolation, but by working together in research groups for periods of time.

I have tested the effects of the demic structure of science by studying a particular area of science over a period of thirty years. One aspect of this research concerned the chief outlet for the work of these scientists, *Systematic Zoology*. As first one "school" (or deme) and then another gained ascendancy, members of these demes consciously tried to get people predisposed to their views appointed as editor of this journal. They succeeded but not with the intended effects. The ratio of papers in favor of a particular deme to those critical of it changed through the years but not in conjunction with the allegiance of the editor. The social structure of science turns out to be sufficiently robust that it can overcome the preferences of individual editors (Hull 1983a). I also studied the refereeing process of this journal. For a period of six years, I examined the referees' reports for all the manuscripts submitted to *Systematic Zoology*, distinguishing whenever possible the deme to which the author and referees belonged. When these scientists were broken down broadly into schools, the degree of bias that I had anticipated did not materialize. Only when these authors and referees were classified into very small groups did social alliances make a significant difference. One message of my studies is that crude classifications of scientists into groups is not good enough. If the effects of the social structure of science are to be discerned, scientists must be studied in detail, in fact, in so much detail that many workers may be discouraged from initiating such an undertaking (Hull 1988a).

The most prevalent way of grouping scientists is according to subject matter – into physicists, biologists, psychologists, etc., but none of these assemblages has much to do with the social structure of science. These groupings are not only artificial but also too large. All physicists do not form a genuine social group. Even high-energy physicists form too large and disparate an assemblage to function as a genuine group in the ongoing processes of science. When Blau (1978) studied the five hundred or so physicists working in high-energy particle physics in the United States at the time, she discovered that roughly half of these physicists were organized into small "cliques" or research teams, while the rest worked primarily alone. She discovered further that about half of these cliques were organized loosely into a more inclusive invisible college. Her most relevant finding for our purposes is that scientists working in cliques, especially those cliques that were part of this larger invisible college, were the most productive.

In response to criticism of the effect of old boy clubs in the funding of science, the National Academy of Science established a committee to undertake a massive study of funding patterns in the National Science Foundation. The results were puzzling in several respects. For example, one of the questions that the committee addressed was the effect that the prestige of one's department had on one's success in obtaining research funds from the NSF. It turned out that reviewers were more likely to be selected from top-ranked departments, but the reviewers from these departments tended to give lower than average scores to proposals from other top-ranked departments. In general, the prestige of one's department did not correlate significantly with success in getting funded.

The authors of the report that stemmed from this study, Cole, Rubin and Cole (1978), suggested two explanations for this and other unexpected findings – self-selection and lack of consensus. Because competition for research funds is so keen, most scientists do not even submit proposals to the NSF. Of those that do, a large number are turned down several times and cease to apply. As a result, the applicant pool is already skewed toward extremely productive scientists. More important for our purposes is lack of consensus. If science exhibits the demic selection suggested by treating it as a selection process, it should be composed of numerous, relatively isolated groups of scientists who hold significantly different views from each other, and these groups are not coextensive with departmental affiliation. As real as departmental boundaries are for administrative purposes in universities, they have little to do with the structure of science. Most scientists working in the same department are working on problems that are so disparate that they might as well be in different universities or on different continents for all the use they are to each other. The pooling of cognitive resources necessary for research frequently requires scientists working in different departments to become part of the same research group.

Cole, Rubin and Cole (1978) were surprised to discover such differing opinions by referees of the NSF proposals that they were sent to evaluate. The luck of the draw played a much bigger role in determining funding than they had expected. In the test that they ran, they discovered that opposite decisions occurred in 25 percent of the cases when the proposals were sent to different referees (Cole, Cole and Simon 1981). However, if science is a selection process, such lack of consensus is not a weakness but a necessary prerequisite for conceptual growth. Variation is required for selection to occur, but more importantly such variation should be "clumped." Papers and research proposals are not isolated entities but are part of the careers of individual scientists and research groups. Competition occurs among "ideas" in science, but it also occurs among the scientists who hold these ideas. Conceptual selection cannot be treated as selection among disembodied ideas.

The preceding data bear in a general way on treating science as a selection process, but thus far I have said nothing about the detailed models introduced by Kitcher (1992) and Goldman and Shaked (1992). Kitcher's investigation was motivated by the fear that deference to authority, rather than thinking and observing for oneself, might stifle innovation. In this connection, he distinguishes between earned and unearned authority. Earned authority is the credibility that results from the scientist's performance and the opinions of those who know about this performance. Unearned authority stems from being trained by a prominent figure, position in a top-ranked department, etc. Usually, the two should go together, but what if they do not? If the primary locus of research is as narrow socially as I have argued above, then scientists rarely have to resort to unearned authority in deciding which work to ignore, take on faith, or test. In most cases, they will already have sufficient first-hand knowledge of the credibility of the workers in question.

Merton's Matthew effect would seem to imply that unearned authority plays a significant role in science. In his original paper, Merton (1968) suggested that there might be a "ratchet effect" operating in the careers of scientists such that increased recognition leads to increased recognition.[2] All else being equal, scientific authorities receive greater recognition for their work than lesser figures in the field. The two areas in which this effect should be most noticeable are in collaboration and independent multiple discoveries. If a major and a minor figure coauthor a paper, the major figure is likely to receive a disproportionate amount of the credit for this paper, especially if the minor figure stays minor. If a major and minor figure publish the same finding at the same time, the major figure will tend to be given credit for the discovery, especially if the minor figure stays minor. With respect to Kitcher's concerns, authorities should have more authority than more minor figures.

Merton (1968) supported his hypotheses by reference to interviews and data drawn from the history of science. When Cole (1970) gathered more extensive data, he was forced to conclude that the Matthew effect explained very little of the variance in the cases that he investigated. The work of well-known authors received more attention early on and was diffused more rapidly, but good papers end up receiving the same attention, regardless of the prestige of their author. So much authority is attributed to authorities because they produce so much good work, not simply because they are authorities. Hence, unearned authority does not seem to play a very large role in the decisions that scientists make. As mentioned above, Cole, Rubin and Cole (1978) came to the same conclusion with respect to the allocation of funds by the NSF. Cole (1975) did find strong confirmation for the hypothesis that a high proportion of citations are for support. Now we need to know if more prestigious authors are cited for support more frequently than less well-known scientists.

Goldman and Shaked (1992) investigate the effects which credit-seeking and truth-seeking motivations have on truth acquisition. Credit-seeking motivation is fairly easy to operationalize in a way that is reasonably independent of any truth-seeking motivation: scientists sign their papers, negotiate the order of names on multi-authored papers, cite their own work more frequently than the work of anyone else, engage in priority disputes, rush into print when the competition is keen, etc. Some of the preceding might in some way be connected to the acquisition of truth. For example, if you think that your work *is* the best, then you are fostering the acquisition of truth by citing it. However, numerous instances can be given in which the desire for credit has inhibited the acquisition of truth. The search for ways of producing insulin (Bliss 1982) and the discovery of the molecular structure of the pituitary hormone (Wade 1981) are but two examples, but I know of no summary figures that bear on the issue.

Operationalizing truth-seeking motivation in ways that are reasonably independent of credit-seeking motivation is more difficult. What would indicate that a scientist was more interested in truth and credit? One possibility is the running of extensive, careful experiments without regard to the progress being made by other scientists on this particular problem. You will adhere to proper scientific method, regardless of the corners being cut by your competitors. In fact, you refuse even to view them as competitors. Unfortunately, time marches on even for the most idealistic scientists. The only truths worth pursuing are those that are not yet known. Once someone else has published a paper in which the problem on which you have been working for several years is clearly solved, there is little point in your continuing to pursue your research for another ten years, bringing it to the sort of conclusion that you think proper scientific method demands. If you did, no one would publish your results anyway.

Another sign that scientists are interested in truth and not credit is the alacrity with which they admit in no uncertain terms that those who have shown their earlier work to be clearly mistaken are right. Again, I know of no summary data on this question, but for many years I have been keeping a list of such examples, and it is very short. Three recent additions to this list are in connection with the cold fusion debacle. Three laboratories that claimed to have discovered effects that confirmed Pons and Fleischmann's speculations have since publicly recanted (Close 1990). Scientists are quite reluctant to admit that they are mistaken, and this stubbornness sometimes pays off, when it turns out that they were right all along, overwhelming data against them notwithstanding.

The social structure of science is designed to apply constant pressure on scientists to pursue their labors both carefully and intensely. Sometimes the pressure gets so great that truth acquisition suffers, but the other alternative can have just as deleterious effects. The road to hell is paved with good intentions in science as elsewhere. Many social planners have meant well. They really did want to help humanity. However, their lack of success indicates that idealistic motivations alone are far from sufficient. Would truth-motivation alone be sufficient in science? I doubt it. Science is so organized that scientists who have both sorts of motivations are likely to be more successful in discovering truth than those who are motivated by the desire for acquiring credit or truth alone.

CONCLUSION

Science is sufficiently important for us to use the same sort of standards and exert the same amount of effort to understand it as we do in studying the rest of the world in which we live. It is possible that conceptual change in science exhibits the same sort of structure to be found in other selection processes. If so, explicit models that exemplify certain aspects of this process need to be worked out in detail and then tested. Isolated case studies are not considered sufficient evidence in other areas of science. They should not be considered sufficient evidence in the study of science either. Unfortunately, the profession that in the past has been committed to the gathering of the sort of data necessary to test hypotheses about science has recently undergone a significant shift in emphasis. Sociologists of science have abandoned the old "positivistic" view of science in which data are held to influence (though not compel) assent. The sociology of science has been largely displaced by the social construction of knowledge. Science has been replaced by philosophy and discourse analysis. Until enough people trained in the sociology of science become interested once again in testing general views about science, our understanding of science, whether as a selection process or any other process, will remain impressionistic and programmatic.

NOTES

Reprinted with permission from *World Futures* 34 (1992): 67–82.

1. As might be expected, numerous objections have been raised to Pendlebury's study. Most serious is the amount of "marginalia" that the ISI data base includes as "articles." When meeting abstracts, editorials, obituaries, etc. are eliminated the percentage of papers that are not cited within five years decreases but still remains sufficiently high in some areas to be a matter of some concern. The percentage of papers not cited in the natural sciences drops from 47.4 to 22.4, in the social sciences from 74.7 to 48, and in the humanities from 98 to 93.1 (Pendlebury 1991). Strangely, work in the history and philosophy of science is counted part of the humanities, and yet only 29.2% of the papers in history and philosophy of science remain uncited within the first five years after publication against 36.7% in physics! Something is wrong somewhere.

2. H. J. Muller (1964) pointed out a similar ratchet effect in the accumulation of mutations in asexual organisms.

9

Testing Philosophical Claims about Science

With respect to the role of evidence in testing statements both *within* science and *about* science, four combinations are possible. Logical empiricists such as Hempel (1966) insist that evidence plays a crucial role in the sort of testing that goes on in science. In their own discussions of science, logical empiricists also include occasional examples drawn from science, both current and past, but these examples function only as illustrations of the points that they are making about science, not as tests. A common view among philosophers of science, and not just logical empiricists, is that no connections exist between what scientists actually do and the sorts of claims that philosophers make about science. Even if no scientist ever explained anything by deriving it from a law of nature, the covering-law model of scientific explanation would remain untouched. Deduction is deduction, and nothing about the conduct of science can touch that. According to these philosophers, evidence may play a crucial role in testing the empirical claims that scientists make about the natural world but no role whatsoever in testing the sort of meta-level claims that philosophers of science make about science. Such philosophical claims must be supported in some other way.

According to one prevalent reading, Kuhn (1962) advocates the opposite position with respect to the role of evidence in science and the study of science respectively. To the extent that evidence plays any role whatsoever in scientists' choosing between different paradigms, it is never decisive. However, Kuhn urges a greater role for the history of science in the choice between different philosophies of science. Incommensurability between scientific theories precludes a decisive role for evidence with respect to theory choice within science, but the even greater incommensurability that characterizes different philoso-

185

phies of science somehow does not preclude a decisive role for the history of science as evidence in choosing between these meta-level theories.

Some of Kuhn's social constructivist disciples have carried his position to even more extreme lengths. As Collins (1981a, p. 218) sees it, advocates of the radical program in the sociology of knowledge "must treat the natural world as though it in no way constrains what is believed to be." However, sociologists of science should "treat the social world as real, and as something about which we can have sound data" (Collins 1981a, p. 217). Other social constructivists have pursued what they take to be Kuhn's views to their logical conclusion – total relativism. Evidence plays no role in either science or meta-level investigations of science. Just as the natural world does not constrain our interpretations of the natural world, texts do not constrain our interpretations of these texts (Woolgar 1988).

The fourth alternative is that evidence can play a significant role in both science and the study of science. Evidence is not easily brought to bear on general claims in science. Showing the relevance of evidence in testing meta-level claims is even more difficult, especially since students of science make very different sorts of claims about science. For example, philosophical claims about the adequacy of operational definitions are different in kind from sociological claims about the disproportionate effect that the work of a very few scientists has on the course of science. However, those of us who see a need for testing meta-level claims made by students of science insist that even the most philosophical claims about science can be interpreted so that evidence can be brought to bear on them, albeit sometimes quite indirectly. Certainly the sorts of claims made by sociologists of science can be tested empirically.

1   TESTING META-LEVEL CLAIMS

If the theory-ladenness of even the most observational of terms really does pose the insurmountable problems in choosing between alternative theories that Kuhnians claim, then we should see the results of this incommensurability in science, both past and present. If different paradigms are incommensurable, then some fairly obvious conclusions follow about how successfully scientists who hold the same and different paradigms can communicate with each other. Scientists do, not infrequently, talk past each other, but does this failure in communication covary with adherence to different paradigms?

One would think, given the huge literature on the subject of incommensurability, that numerous authors would have attempted to test this apparent implication of Kuhn's thesis in a systematic way. Lots of case studies have been presented, some showing how incommensurability did pose a serious problem in communication, some showing how it did not. If the correlation between incommensurability and communication is taken to be universal and the Popperian asymmetry between verification and falsification is taken seriously, then a single contrary instance should refute this meta-level claim. Several refuting instances have been presented.

For instance, Ruse (1979) has detailed the controversy over the age of the earth that took place in the second half of the 19th century between evolutionists and geologists, on the one hand, and physicists, on the other hand. Evolutionists and geologists thought that the earth was extremely old, while physicists insisted that it was relatively young. Neither side was able to come up with extremely precise figures, but the physicists thought that the earth has been around for at least twenty-five million years, possibly as long as a hundred million years, while the evolutionists and geologists insisted that it had to be much older – hundreds of millions of years.

If any two groups of scientists ever held incommensurable paradigms, the Darwinians and the Kelvinians did. They deployed different symbolic generalizations, employed different methodologies, shared different professional values, and most importantly extrapolated from very different exemplars. However, in spite of all these differences, these two groups of scientists were able to disagree with each other just fine. Perhaps they meant something slightly different by "age" or the "earth," but such slight differences in meaning were overridden by the magnitude of the differences in the age of the earth implied by these two paradigms. When Darwin estimated that the denudation of a single stratum took 250 million years and the physicists had a hard time coming up with that figure for the entire duration of the earth, a contradiction clearly existed, incommensurability notwithstanding.

Although case studies are in principle sufficient to *refute* a general thesis, in practice they rarely do so. Too many objections can be raised to their relevance, applicability, construction, execution, etc. They are not even in principle sufficient to *confirm* a general thesis. Rarely, however, are theses about science presented in a universal form.

Usually they are hedged here and there. For example, sometimes Collins (1981a, p. 218) portrays the radical program in the sociology of knowledge as requiring that the "natural world in no way constrains what is believed to be." At other times, it requires only that the "natural world has a small or non-existent role in the construction of scientific knowledge" (Collins 1981b, p. 3). Systematic, preferably quantitative, studies are required to test claims such as these (Cole 1992).

No one to my knowledge has even attempted such a study with respect to the effects of incommensurability on success in communication. In my own research, I have studied these effects in a semisystematic way (Hull 1988a). I found no clear correlation. Confusion was as common within groups of scientists holding the same paradigm as between groups holding different paradigms. I realize that this lack of correspondence is impossible, but as far as I can tell, it is actual. Either there are so many other causes for failure to communicate successfully in addition to incommensurability that they swamp the effects of incommensurability, or else incommensurability does not present the insoluble problems that holistic semantic theories seem to imply that they should.

2   IDEALIZATIONS

One obvious response to the above comments is that the connection between philosophical analyses and science is not as simple as I make it out to be. Philosophers discussing the problem of incommensurability are not talking about science as it is practiced but about idealizations of their own construction. Idealizations play legitimate roles in science. Perhaps they play equally legitimate roles in our analyses of science.

In order for two theories actually to contradict each other, they must be presented in complete, totally precise, possibly axiomatized form with all meanings sharpened to a fine point by sufficient conceptual analysis. Only then can the two theories be shown to be incommensurable. Scientists do not present their theories in such an ideal form. Nor do they evince any interest in doing so. At one time philosophers set themselves the task of producing such ideal versions of scientific theories. However, such undertakings are now decidedly out of favor. Instead, we are asked to consider problems that *would* arise *if* we had two perfectly formulated theories. For such ideal formulations, incommensurability would be a problem.

Scientists test ideal laws within science by seeing how real systems behave as they approach the ideal. Some inclined planes exhibit less friction than others. When actual surfaces are ordered according to their degree of friction, the results approach the ideal. Even if incommensurability characterizes only those theories that are completely and perfectly formulated, actual theories can be ordered to see if incommensurability becomes a greater problem as this ideal is approached. With respect to the theories that scientists actually produce, deciding which observation statements follow from these theories and which do not is far from easy. All sorts of approximations and simplifications have to be introduced, just the sorts of approximations and simplifications needed to derive commensurable observation statements from different theories. As a result, attempts to test one theory in isolation do not look all that different from attempts to test two theories by inferring incompatible and, hence, commensurable observation statements from each. As science proceeds and theories in a particular area become better formulated, the issue of incommensurability should become even more prominent. So far no one has attempted a study to see if incommensurability becomes increasingly more evident as scientists make their theories increasingly precise.

The issue is, as before, the testing of meta-level claims. Given a holistic semantic theory, incommensurability follows automatically. No evidence about the actual course of science is in the least relevant. On this view, semantic theories have nothing more to do with communication than the covering-law model of scientific explanation has to do with how scientists explain natural phenomena. Claims made by philosophers of science may sound as if they are about science and can be tested by reference to science, but in point of fact they express philosophical theses so abstruse that nothing so crude as evidence can be brought to bear on them. If this is the position that philosophers adopt, then detailed case studies are just so much deceptive window dressing. If all case studies are supposed to do is to illustrate a particular point, then brief gestures or silly science fiction examples will do. All the effort needed to set out real examples in all their complexity is wasted. Loading a philosophical discussion with detailed history of science may fool the unsuspecting reader into thinking that the author is talking about science, but that is all.

3 STUDYING SCIENCE

Those of us who are not inclined to take the a priori route are still left with plenty of problems. Bringing history of science to bear on general claims about science is extremely difficult. One of these difficulties is a meta-level version of a problem raised by philosophers in the context of science itself – theory-ladenness. Within science, observation terms are laden with the very theories that these observation statements are meant to test. If you approach the relevant data from the perspective of a particular theory, e.g., Darwinian evolution, which data seem relevant and how you construe these data will be strongly colored by your beliefs about the evolutionary process. Hence, you should not be surprised when your observations *support* your theory. However, you should be surprised when they *contradict* it, but contrary to a priori expectations, sometimes they do. No matter how strongly one's general views color one's estimations of data, sometimes these data can challenge the very theories in which they are generated. It should be impossible, but once again it is actual.

For example, T. H. Morgan's investigation of fruit flies eventually led him to abandon nearly every basic belief that he designed his experiments to support. He also reported no conversion experience as he abandoned one paradigm for another. Rather, he painfully modified one belief after another as the experiments that he and his students ran forced him to. Certainly career interests influenced Morgan the way that they influence all scientists, but if "interests" of a broader sort played a significant causal role, it is far from apparent. Morgan and his co-workers in the fly room were middle-class Euromales before they began their investigations; they remained so afterwards.

As Lakatos (1971) has pointed out, we are confronted by parallel problems in describing the course of science. We come to the study of history of science with all sorts of beliefs about history and science. These beliefs are only half-formulated and most not even explicit. To make matters worse, they usually do not deserve to be called a "theory" of history. Hence, their influence on the "data" that are generated are likely to be even more pervasive and elusive than the parallel situation in science. As Richards (1993) points out in his paper, most practicing historians of science are crude inductivists when it comes to their own work, even those historians who reject inductivism as adequate for the practice of science itself.

The influence of the general beliefs held by historians on the stories that they tell are obvious. Recently, Columbus bashing has become fashionable. Did Columbus really discover America? After all, human beings already inhabited the continent when Columbus arrived. They got there long before via the Bering Strait land bridge. And Columbus may not have been even the first Old World person to set foot in the New World. Perhaps some Norseman or Viking may have made it there first. In addition, Columbus was not the only European on his boat. He was not the first to sight land. One of his men did. How come none of his crew members get any credit? They discovered America too. To make matters worse, Columbus did not think that he was discovering America. He thought he was arriving at the eastern shores of the Indies. Nor did he term his discovery "America." This name was coined much later, and on and on.

As trendy as the preceding discussion may sound, it introduces no problems not already familiar to historians of science. Did Mendel really discover Mendelian genetics? One can find numerous examples of three-to-one ratios in the works of his predecessors. Besides, Mendel did not think that he was discovering the laws of genetics, let alone Mendelian genetics. He thought he was investigating speciation by means of hybridization. William Bateson at the turn of the century was the one who was primarily responsible for transmuting Mendel's observations on peas into the science of genetics and making Mendel its patron saint.

As critical as I am of the general philosophical views of the social constructivists, they have forced us to see the bias that is introduced in the study of science by an overemphasis on "great men." For example, Desmond (1989) has shown how different the impact of Geoffroy St. Hilaire on Victorian science looks when viewed from the perspective of ordinary anatomists in medical schools rather than from the perspective of such big guns as Lyell, Owen and Darwin. The story of natural history in Victorian Britain reads very differently when it is written to include lesser lights along with the major figures, as different as the discovery of America looks from the perspective of Columbus's crew.

History of science cannot be written from no perspective whatsoever. It also cannot be written from all possible perspectives. All anyone can do is to be explicit about the perspective that one brings to a particular study. If our meta-level paradigms were so powerful that no observation couched in them could possibly refute them, then we would be in real trouble, but as in the case of science, students of science come up with observations about science that do not fit neatly into their own belief systems. For example, Popperians have attempted to test (or possibly only illustrate) Popper's views by recourse to the history of science. As biased as they may have been in favor of Popper's worldview, they were not always able to make the stories come out right – at least not without massive rational reconstruction. Bringing evidence to bear on ordinary empirical claims is not easy. All the problems that beset such efforts are only magnified in testing meta-level claims. Even so, these problems are not so hopeless that they cannot be overcome, if only we actually try.

4 OPERATIONALIZING IN THE STUDY OF SCIENCE

In science theoretical claims have to be operationalized in order to be tested. Such operationalizations require reduction in scope as well as the introduction of rough approximations and particularizations. For example, biologists have long assumed that dinosaurs, like extant reptiles, were cold-blooded. When the suggestion was made that they might have been warm-blooded, biologists had to decide how such an hypothesis might be tested. Among extant predators, the ratio of predator to prey among warm-blooded predators is 1:50, while this same ratio among cold-blooded predators is 1:5. Since the difference is so great, just possibly the fossil record is good enough to choose between these two alternatives.

Of course, without even pausing to breathe, any red-blooded biologist can think of indefinitely many objections to the operationalizations required for this test (and some already have), but as sceptical as scientists are, they are much more tolerant of imprecision and possible error than philosophers are. The sort of argumentation that occurs in science does not come close to the extremely high standards that we set for ourselves. For example, using citations to gauge the impact of a paper on a particular area of science is a crude measure of importance, but it is the sort of operationalization common in science. I am sure that most philosophers would find the operationalizations that I used in assessing "success" in communication hopelessly inadequate, but it is this attitude that precludes philosophers from testing their beliefs about science by reference to science.

A common view expressed by philosophers about science is that the meanings of theoretical terms emerge only in the context of testing. Scientists cannot know in advance of their empirical investigations what they mean by their more general concepts. Meanings change as knowledge advances. The same should hold in our study of science if we are to provide theoretical definitions for our meta-level terms. We cannot possibly know what "testing," "experiment," and even "science" mean in advance of any and all empirical investigations. If we propose to test our knowledge of such general concepts as testing, we have to be willing to accept for the purposes of a particular study operationalizations that we all agree are crude, not good enough by half, etc. But this is the only way to improve upon these operationalizations and, hence, our understanding of these general concepts. Conceptual analyses by intelligent ignoramuses can get us only so far.

Explications of theoretical terms in science by philosophers require our entering into the scientific process. If we are to assess the adequacy of a particular analysis of "species" or "gene," we have to persuade the relevant biologists to incorporate these conceptions into their own work and see what happens. That way they become theoretical definitions rather than conceptual analyses (Milliken 1984, Neander 1991). If conceiving of species as spatiotemporally restricted and located historical entities helps biologists improve evolutionary theory, then this conception must have something going for it. Such a procedure has the added virtue of allowing philosophers to support their theses by pointing to scientific usage.

What we need in science studies is theoretical definitions, not conceptual analyses, and theoretical definitions require theories. Although grand theories about the nature of science are currently out of fashion, I think that we need to rehabilitate them. We need to construct theories about science the way that scientists construct theories about fluids, gene flow and continental drift. To construct such theories, we need data, and our only source of data is the study of science, past and present. These histories will be theory-laden. So? If scientists can use the data generated in the context of theories with which they disagree, why can't we? I do not agree with all the general views about science held by Rudwick (1985) and Desmond (1989), but I have no trouble using their histories to test my own general views about science. The stories that historians tell are theory-laden but not so theory-laden as to be useless. Needless to say, I have my own grand theory of science (Hull 1988a).

5   NORMATIVE CLAIMS ABOUT SCIENCE

Not all scientific claims are simply descriptive. Some are also nomic. As difficult as it is to set out criteria that mark this distinction, I am still old-fashioned enough to think that there are laws of nature and that a continuing goal of science is to discover these laws. Similarly, I hope that not all of the claims made by those of us who study science are going to be simply descriptive. Some I hope will turn out to be analogous to laws of nature – laws about the scientific production of knowledge. If the distinction between descriptive and nomic claims within science is so difficult to set out, the parallel distinction at the meta-level will surely be even more difficult. The notion of meta-nomic necessity may sound overly ambitious to many people, but it is one of the prerequisites for a successful empirical theory of knowledge acquisition in science.

Philosophers have also traditionally expressed normative claims about science. In general, such prescriptions are extremely difficult to test. One suggestion is to convince groups of scientists. Have them adopt one's views about how science *should* be conducted and see what happens. If science in such areas immediately grinds to a halt, then possibly something is wrong with one's normative claims. Conversely, if those scientists who adopt your views are even more successful in attaining their epistemic goals, then possibly there is something to be said for these norms. For example, scientists do not spend much time precisely replicating the work of other scientists. They adopt the results that support their own views without testing. They tend to reserve testing for those results that threaten their own findings, and these tests are rarely exact replications – whatever that might mean. For some, this lack of precise replication may seem a fault, as if scientists are somehow falling short of proper scientific conduct. Scientists *should* replicate all results before using them. However, I strongly suspect that if enough scientists adopted such a prescription, scientific progress would be sharply curtailed.

In emphasizing how important it is for scientists to understand and incorporate the views set out by students of science into their own work, I am not committing myself to the position that scientists are the final arbiters with respect to matters about science as well as within science. To the contrary, I think that philosophers are right about the fundamental inadequacy of operationalism as a philosophical thesis, regardless of what certain behavioral scientists may think on this score. They are mistaken and could easily discover their mistake by reading a few well-chosen papers on the subject. These papers need not be written by philosophers, but the issues will nevertheless be philosophical. However, I do think that the proof of the pudding is in the eating, and only if scientists come to incorporate explicitly formulated meta-level beliefs about science in their own work can we ever hope to see what effects they have on science.

NOTE

Reprinted with permission from *PSA 1992*, vol. 2, edited by D. Hull, M. Forbes, and K. Okruhlik (East Lansing, MI: Philosophy of Science Association, 1993), pp. 468–75. © 1993 by the Philosophy of Science Association. All rights reserved.

Thanks are owed to Kim Sterelny and Todd Grantham for reading and commenting on an early draft of this paper.

10

That Just Don't Sound Right

A Plea for Real Examples

> If thought experiments in science are ultimately supposed to
> increase our understanding of the world by contributing to the
> process of changing the framework of concepts in terms of which
> we think about the world, then thought experiments in philoso-
> phy of science should be directed towards changing and clarify-
> ing the concepts that pertain to science itself.
>
> – Forge, "Thought Experiments in the Philosophy
> of Physical Science"

Although Kuhn published a paper in 1964 arguing for an important
function for thought experiments in science, his fellow philosophers did
not find this topic all that interesting until recently. Of the papers, books
and anthologies that have appeared in the last half dozen years, most
deal with thought experiments in science, primarily physics (Gooding
1990, 1993; Brown 1991, 1993; Nersessian 1993; Hacking 1993).
However, a few also deal with thought experiments in philosophy, especially the philosophy of science (Wilkes 1988; Hull 1989a; Horowitz and
Massey 1991; Sorenson 1992). To my way of thinking, philosophers
have done an excellent job investigating the role and justification of
thought experiments in *science*. We have been much less successful in
setting out the relevant issues with respect to *philosophy*.

Hypothetical reasoning has been part of Western thought since the
pre-Socratics (Rescher 1991). More narrowly, Western intellectuals frequently make use of made-up examples. Some of these fictitious examples could occur but have not. Others are so described that they could
not possibly occur. Some examples of hypothetical reasoning also
count as thought "experiments" in a narrow sense. The authors cited
above tend to concentrate on thought experiments, narrowly construed. In this essay I examine the role of fictitious examples in both

science and philosophy, specifically the philosophy of science, without limiting myself to thought experiments.

Recently a few philosophers, while acknowledging the positive role that thought experiments have played in certain well-articulated areas of science, logic and mathematics, question the legitimacy of appeals to thought experiments and science fiction examples in philosophy. According to Wilkes:

> Personal Identity has been the stamping-ground for bizarre, entertaining, confusing, and inconclusive thought experiments. To my mind, these alluring fictions have led discussion off on the wrong tracks; moreover, since they rely heavily on imagination and intuition, they lead to no solid or agreed conclusions, since intuitions vary and imaginations fail. What is more, I do not think that we need them, since there are so many actual puzzle-cases which defy the *imagination*, but which we none the less have to accept as facts. (1988, p. vii)

Fictitious examples in philosophy have led to wild speculations that are all but impossible to evaluate, but such examples can also have the opposite effect. In both philosophy and science, thought experiments can inhibit innovation, relying as they often do on well-entrenched intuitions. From the perspective of old-think, new ideas just don't sound right. Another problem with thought experiments, especially of the cuter variety, is that they are incoherent. Too often, in the face of the ubiquitous question "What would you say if . . . ?" I am forced to respond that I don't have the slightest idea (Fodor 1964). Many people seem to be able to picture bizarre hypothetical situations with amazing ease. Others, and I am among them, are unable to replicate this enthusiasm because we simply can't conceive of the situations described. Either we are psychologically deficient – sort of philosophical dyslexics – or else we have very different standards of conceivability. Just flitting before the mind's eye is not good enough. As Goldman complains, "Some of the most recent discussion of the analysis of knowledge has taken on a glass-beads-game quality – counterexamples dazzling in their originality, designed to capture intuitions few of us recognize as such, generating additional criteria for knowledge so complex in their application as to be of no use in deciding when and what we know" (1988, p. 19).

If, as Horowitz and Massey (1991) argue, thought experiments in the broadest sense have supplanted meaning analysis in analytic philosophy, then one would think that those philosophers who appeal to thought experiments would have set out in great detail standards for their use in philosophy. With the exception of Sorenson (1992), they have not. As Massey asks in connection with thought experiments such as Putnam's Twin-Earth example:

> How do *I* know whether you have been successful? How for that matter do *you* know whether you have been successful? Well, if you feel satisfied with your conception, if it seems to you that you have managed to conceive such a state of affairs, you simply rest your case. And if it seems to me that you have succeeded, perhaps because of profiting from your tutelage I seem now able to conceive it myself, I capitulate. Either I abandon my thesis or, like Hume, I dismiss your counterexample as too esoteric to do serious damage. There is, of course, a third alternative, one surprisingly rarely taken, namely, to dismiss your conception as somehow bogus or counterfeit. (1991, p. 292)

Critics of the use of science fiction examples in philosophy find that such appeals are too "free and easy." In a sort of Protestant rebuff, they reason that anything *that* easy cannot be of much value, and I agree. The sort of science fiction examples to which analytic philosophers seem peculiarly prone have three glaring deficiencies. First, they lack sufficient detail. We are rarely told enough to understand the state of affairs being described (Dancy 1985). Second, they lack a theoretical context to enable us to fill out the description on our own (Wilkes 1988, pp. 7, 21, 44–5; Massey 1991, p. 294; Horowitz 1991, p. 306; Irvine 1991, p. 159). In real examples, the background knowledge is already known and widely shared. If some item in the background knowledge is brought into question, it is there to be investigated. In made-up examples, no such background knowledge exists. Real examples pose as fundamental problems as do made-up examples. They also have the advantage of presenting ways by which these problems can be solved. Third, in using such examples we trade on a notion of conceivability that remains all but unexplicated.

One way around this lack of detail is to avoid making one's examples too outlandish, too different from actual states of affairs (Wilkes 1988, p. 46; Forge 1991, p. 216; Gale 1991, p. 23). As Gooding remarks, legitimate thought experiments "posit a world – neither so familiar as to foreclose change nor so strange as to provide no footholds or handles on reality" (1993, p. 283). Thought experiments "persuade when there is enough strangeness to disturb and enough familiarity to be accessible." Thus, in our hypothetical examples, we can trade off what we already know about related real examples. In the limiting case, our hypothetical examples are not in the least science fiction. For example, the Ship of Theseus is an ordinary ship. The only strange thing about it is that it has to be rebuilt while it stays afloat at sea.

In scientific thought experiments, we are asked to countenance the modification of a single feature of the situation while everything else remains the same. Such examples conform to the "laws we know and trust." Put differently, "the 'possible world' is *our* world, the world described by our sciences, except for one distinguishing difference" (Wilkes 1988, p. 8).

One problem with this interpretation of hypothetical examples is that certain areas of science are tightly organized. Modifications ramify. As John Herschel remarked in his early work on the study of natural philosophy:

> The liberty of speculation which we possess in the domains of theory is not like the wild licence of the slave broke loose from his fetters, but rather like that of the freeman who has learned the lessons of self-restraint in the school of just subordination. (1841, p. 190)

Present-day physicists agree with Herschel about constraints on the liberty of speculation. As Feynman, Leighton, and Sands remark:

> The whole question of imagination in science is often misunderstood by people in other disciplines. They try to test our imagination in the following way. They say, "Here is a picture of some people in a situation. What do you imagine will happen next?" When we say, "I can't imagine," they may think we have a weak imagination. They overlook the fact that whatever we are *allowed* to imagine in science must be *consistent with everything else we know*: that the electric fields and the waves we talk about are not just some happy thoughts which we are free to make as we wish, but ideas which must be consistent with all the laws of physics we know. We can't allow ourselves to seriously imagine things which are obviously in contradiction to the known laws of nature. And so our kind of imagination is quite a difficult game. One has to have the imagination to think of something that has never been seen before, never been heard of before. At the same time the thoughts are restricted in a strait jacket, so to speak, limited by the conditions that come from our knowledge of the way nature really is. The problem of creating something which is new, but which is consistent with everything which has been seen before, is one of extreme difficulty. (1964, 2:20. 10–11)

In the face of all these objections to using hypothetical examples in philosophy, philosophers might well be led to specify their examples more fully, supplying more detail and indicating the context in which these examples are to function. If no such context exists, philosophers need to construct one. Although these requirements might sound excessive to philosophers, authors of fiction have been fulfilling them for centuries. If Jane Austen can do it, so can Hilary Putnam.

Another alternative is to use real examples. In his "defense" of thought experiments in philosophy, Sorenson complains that too often philosophers use thought experiments as a mental crutch:

> Instead of taking the extra time to find a robust actual case, they fall back on anemic hypotheticals. Thought experiment is an easy way of getting short-term results. Philosophers already know how to do it, it's cheap, it's quick, it's accepted. (1992, p. 256)

What is so wrong with philosophers using actual cases? One answer is that real examples can be difficult to explain, and in the midst of all the empirical detail, the philosophical points at issue can get lost. Goodman uses precisely this justification for resorting to grue and bleen ([1954] 1983, p. xix). He thinks that commonplace and trivial illustrations "attract the least interest to themselves" and hence are "least likely to divert attention from the problem or principle being explained." Perhaps so, but his own example of grue emeralds is a major exception to this generalization. It has diverted massive attention to itself. For forty years, philosophers have discussed what a world would be like in which the predicate "grue" would be in some sense "natural." In addition, if the calls for greater detail in fictitious examples are heeded, then hypothetical examples are likely to become as complex as their real counterparts. To make matters worse, no one has suggested how we are to go about adding details to fictitious examples.

One reason philosophers give for resorting to made-up examples is that the real world is too limiting. We need to go beyond the actual world to test and possibly expand our concepts. We need to conjecture possible worlds. I agree that if knowledge is to increase, we have to go beyond what we already know, but I have one suggestion for my fellow philosophers – *proceed more slowly from the real world to possible worlds. Fully exploit the world as we know it before conjuring up exotic possible worlds.* This is the strategy that Wilkes recommends for the huge literature on personal identity. According to Wilkes, "basing our arguments on actual cases allows us to check our imagination against the facts, and our intuitions get strengthened and rendered more trust-worthy" (1988, p. 48). Once armed with these strengthened and trust-worthy intuitions, we can then risk the construction of hypothetical situations if need be.

In biology, the area of science with which I am most familiar, biologists rarely go beyond the real world. They already have such long lists of bizarre actual counterexamples that they rarely feel the need to make up additional hypothetical problems. (Although Lennox 1991 has shown that Darwin resorted to thought experiments to argue for the possibility of species evolving, he also inundated his readers with real examples.) For instance, species in sexual organisms are determined by reproductive gaps. If two populations of organisms breed freely with each other when they come in contact and produce fertile offspring, then they belong to the same species. However, there are two species of fish that fulfill this criterion but are nevertheless considered separate species. When they meet, they breed freely with each other and produce offspring that are both fertile and very vigorous, so vigorous that they rapidly out-compete the organisms of the original species. When the last male belonging to the parental species dies, all reproduction ceases because all the hybrids are female. Even though these two species mate freely and produce fertile offspring, they are nevertheless reproductively isolated. It takes just a few generations before this reproductive isolation manifests itself. The supply of similar examples is inexhaustible.

Starting with real examples has the added advantage that one need not have a general theory of conceivability in order to work with them. A major reason philosophers resort to conceivability is to get at possibility. According to Massey's thesis of facile conception, casually alleged conceivability is invoked to establish possibility, and casually alleged inconceivability is invoked to demonstrate impossibility (1991, p. 291). Too often, the decisions that philosophers make rest heavily on intuitions about what sounds right. For example, logical empiricists have long attempted to distinguish between genuine laws of nature and accidentally true universal statements. Although many philosophers think that this attempt is misplaced, there does seem to be an intuitive difference, for example, between the claim that all species evolve through natural selection and that all terrestrial organisms use levo rather than dextro amino acids. The former is fundamental to the evolutionary process; the latter is most likely the result of an historical accident.

One suggestion for explicating this felt difference is by means of counterfactual conditionals. So the story goes, laws can support counterfactual conditionals while accidentally true universal generalizations cannot. If *A* were the case, then *B* would be the case. *A* is not a species, but if it were, it would evolve through natural selection. Laws also support subjunctive conditionals. If *A* should come to pass, then so would *B*. If a species becomes genetically homogeneous, then it will go extinct. The trouble is that these examples get us nowhere. *The intuitions that lead us to accept counterfactual and subjunctive conditionals are precisely the intuitions that lead us to distinguish between genuine laws of nature and accidentally true universals in the first place.*

One virtue of real examples is that they allow us to go directly to possibility without detouring through conceivability. Anything that is actual had better be possible. Whether or not everything that is actual is also conceivable depends on one's general theory of conceivability – a theory that we currently do not possess even in a rudimentary form. As Horowitz and Massey argue:

> Philosophers should muster enough intellectual integrity to eschew conceivability arguments that fail to measure up to the standards that have been developed piecemeal in science. Second, philosophers ought not to rest content with piecemeal developments but should instead turn their talents to the construction of a general theory of conceivability. (1991, p. 23)

Such a general theory of conceivability is likely to be a long time in coming. In the interim, philosophers can get around all the problems generated by a largely tacit notion of conceivability by exploiting real examples to their fullest potential.

Although many real examples are as weird and counterintuitive as the sorts of examples that philosophers make up, they have the virtue of always occurring in a particular context against a wealth of background knowledge – knowledge that can be made explicit when the occasion demands. Although sometimes background knowledge is found wanting when we appeal to it in real examples, at least partially explicit standards exist that stipulate how we should go about supplementing it. As numerous authors have observed, in philosophy such standards are notable for their absence. Typically the examples preferred by philosophers "take us too far from the actual world, and from the 'other things' that hold roughly 'equal' here. . . . This means that we are left with no clue as to what has been varied in thought and what left (supposedly) untouched" (Wilkes 1988, p. 45).

What if another planet existed that was identical to Earth in every respect except that the clear, tasteless, colorless liquid that covers so much of the earth's surface did not have the chemical constitution of water – H_2O? To avoid problems that knowledge of the molecular structure of these two substances introduces, Putnam (1981) assumes that the inhabitants of both planets are just ignorant enough to be oblivious to the chemical and molecular differences between these two substances and the possible ramifications of these differences. The former inhabitants of these two worlds to one side, how about *Putnam's readers now*? Perhaps earlier generations on these two planets were unaware of differences between the two substances that they call "water," but some of us at least know what the chemical structure of water is as well as some general features of the relation between chemical structure and both macro and micro properties. I can conceive of people who cannot tell all sorts of substances apart. I need not conjecture such people on different planets. We have plenty here on earth. But I cannot conceive of two different chemicals having *all* the same characteristics, both macro and micro. Even the rare earths and inert gases exhibit some differences in their salient features. Water also plays very particular and precise roles in the lives of all terrestrial organisms. Life on Earth is a very complex, highly interdependent system. As such it is very sensitive to the slightest changes, especially in something as important to metabolism as water.

"But couldn't you envisage a system in which all the ramifications of 'water' having a different chemical makeup are fully worked out?" Putnam claims that it is "easy" to modify his example so as to avoid these objections (1981, p. 23). I disagree. Perhaps hundreds of scientists working for several decades might work out the details of such a system, but I do not know what it would be like for them to conjure up such a system in the blink of an eye, let alone my doing so all by myself in the space of my reading about Twin Earth (Fodor 1964). "But didn't something flit across your mind?" Something, but not much. Not enough to justify a philosophical analysis of reference.

How about brains in a vat? Mightn't I be a brain in a vat with my nerves connected to a computer presided over by an evil neuroscientist who stimulates my grey matter to maximize false beliefs? Couldn't we all be brains in vats, including the neuroscientist himself?

This time Putnam (1981) introduces his example to show that, contrary to a lot of people's intuitions, no one can really think that he or she is a brain in a vat. Such a belief is "self-refuting" and "incoherent" because the preconditions of reference are absent, that is, *if* we accept Putnam's causal theory of reference. Sorenson disagrees. "How would things seem under such conditions? Just as they do seem! We know all too well what it is like to be a brain in a vat" (1992, p. 7). Hence, we can have legitimate doubt about any of our beliefs (see also Collier 1990). Once again, I must be missing some widespread ability because, quite independent of any causal theory of reference, I do not know what it would be like to be a brain in a vat.

Norton (1993) claims that all thought experiments can be recast as arguments, either deductive or inductive. Within this context, he goes on to claim that the picturesque details in descriptions of thought experiments give them an experimental flavor but are strictly irrelevant to the argument. Nersessian (1993, p. 296) disagrees. These "details usually serve to reinforce crucial aspects of the experiment" (see also Collier 1990). I am of two minds on this issue. If the examples that philosophers use are intended only as expository devices of no intrinsic importance, then I have no objection to them. They might help some readers see the point; others not. Since nothing much rides on them, a brief characterization is good enough. But too often the examples take over, becoming the chief topic of discussion.[1] They are what persuade the reader; not the arguments.

When the examples become the issue, then I for one would like to see more detail, not less. In such situations, I would like to see us take our examples seriously. However, I would much prefer that we begin with real examples, turning to hypothetical examples only when we run out of real ones. Even so, I agree with Norton (1993) that the underlying argument, not the "picturesque" part of the example, is what should carry conviction. In spite of the significant role that Maxwell's demon plays in this literature, Hacking (1993, p. 302) is forced to conclude that the "demon does not, for me, prove even the possibility of anything."

Because fictitious examples play a central role in analytic philosophy, my objections to them and my preference for real examples are likely to make analytic philosophers a bit defensive. How are we to do analytic philosophy without science fiction examples, the more exotic the better? If we deny ourselves appeals to such things as Twin Earth, brains in a vat, violinists hooked up to us while we sleep, etc., all is lost.

204

We will be struck dumb. But are made-up examples really central to analytic philosophy? In thirty years of publishing, I have never attempted to clarify a concept or support a position by reference to fictitious example, except when I have been forced to respond reluctantly to the examples made up by others to criticize my views (Hull 1981a, 1991). Either I am not engaged in analytic philosophy or else use of such examples is not essential to this way of doing philosophy.

In this essay I illustrate the virtues of real examples by reference to (what else?) two real examples: one in science (the species concept), the other in philosophy (natural kinds). Although both the terms and the conceptions have varied considerably through the ages, biological species have been viewed as paradigm natural kinds for over two thousand years, even after Darwin convinced us that species evolve. For over twenty years, Michael Ghiselin (1974) and I (Hull 1976) have argued that insufficient attention has been paid to the implications of Darwin's modification of the species concept. If species are the things that evolve, then they cannot be spatiotemporally unrestricted kinds but must be construed as spatiotemporally restricted particulars (or individuals).[2] Like individual organisms, particular species are space-time worms. In the interim, a substantial number of biologists and philosophers have been persuaded that there is something to this bizarre idea and have incorporated it in their own research.

In the following two sections, I discuss the species concept and natural kinds in the context of evolutionary biology and then turn to some exotic examples that have been introduced to "clarify" these notions. After contemplating these examples, the reader can decide whether they really help or simply introduce a host of extraneous considerations that do nothing but cloud the issues. The issue is *not* whether I am right about species being in some sense "individuals" or even whether the distinction between individuals and kinds can bear up under philosophical analysis but the function of real versus fictitious examples in helping to decide these questions.

BIOLOGICAL SPECIES

From Aristotle to the present, people in all cultures have conceived of biological species as kinds possessing core characteristics – a set of characteristics that all and only members of a particular species possess, a set of characteristics that make a species what it is – an occasional abnormal monster notwithstanding (Atran 1990). The trouble is that, as the biogeographic distributions of species were traced, more and more individuals had to be dismissed as monsters. Species frequently vary throughout their geographic range. Any attempt to subdivide such continua by means of sets of characteristics results in as many abnormal as normal individuals. In addition, some species exist in which males and females differ from each other so dramatically that any attempt to group them according to character covariation results in their belonging to separate species. The same can be said for various castes in social insects and clonal forms. The various organisms that make up a Portuguese man-of-war are phenotypically very different, yet they all belong to the same species.

In response to such character covariation, taxonomists have resorted to what Wittgenstein came to term "family resemblances." Species do not form discrete groups but only "clusters." All that is necessary for organisms to belong to a particular species is for them to possess enough variably weighted characteristics. The general notion of cluster analysis has seemed intuitively satisfying to many scientists and philosophers alike, but actual application is quite another matter. Traits have to be identified and weighted. One clustering algorithm must then be selected amongst the welter of competing computer programs. However, once the amount of labor necessary to use cluster analysis is expended, it works only for *contemporaneous time-slices of those species that exhibit a unimodal distribution* – a single bell curve around a single mean. But many species exhibit multimodal distributions. Which characteristics are "typical" varies from geographic location to geographic location. Averaging this variation to form a single cluster obliterates an important feature of biological species.

Species have a temporal dimension as well. If species are traced through time, they can be seen to split, undergo anagenetic change, go extinct and even merge. As a result, if species are held to have a temporal dimension, then temporal sequences of clusters collapse into a single distribution. The result is one huge smear. In biological evolution something more basic than character distributions is at work. Although natural selection is not the only mechanism that influences the evolution of species, it is primary. Selection is not a matter of picking balls from a jar and tossing them back in. Selection requires the differential build-up of changes through time. Such sequential differential change might occur by a variety of mechanisms, but the only mechanism that has evolved here on Earth requires descent. In a single generation, organisms of certain types produce more offspring than do others. They in turn produce even more offspring relative to their competitors in the next generation, and so on. If species are to evolve by natural selection, then they must be treated as space-time worms. Organisms that belong to a particular species belong to it, not *because* they share the same cluster of characteristics but *because* they are part of the same genealogical network. Because they belong to the same genealogical network, they will tend to possess similar characteristics, but they need not and sometimes do not.

The preceding view of species has proven to be counterintuitive. Some critics have objected that there has to be at least a common core of traits or underlying set of genes that characterize the members of a species or they could not mate with each other (Caplan 1980, 1981; Kitts and Kitts 1979). As plausible as this conviction may seem, it is false. In order for two organisms to mate, they must be sufficiently similar to each other, especially in certain key aspects, but it does not follow from this fact that a single common core of characteristics or genes exists that permits mating. Instead, there are numerous alternative sets of characteristics and/or genes that will do the trick. Any two organisms selected from the same species will have exactly the same alleles at almost every one of its loci, but in different pairs of organisms, *these need not be the same alleles or the same loci*. In addition, interbreeding is not an all-or-nothing affair. Organisms that belong to the same species are *more likely* to mate with each other than with members of other species, but not all members of the same species are interfertile, and sometimes organisms belonging to different species succeed in producing fertile hybrid offspring. It all depends on how much crossbreeding there is. In fact, interspecific hybridization is a common way of producing new species among plants.

Another common objection to the preceding view of species is that this is not how "we" conceive of species. I find this objection misplaced on a host of counts. First, who is the "we" being referred to? Terms are dynamite. Because modern genetics was termed "Mendelian genetics," we tend to think that Mendel had something to do with it. But just as meteors have nothing to do with meteorology, Mendel might not have had that much to do with the science that bears his name. In philosophy there is a movement termed "ordinary language philosophy." Because of its name, one might expect that it has something to do with ordinary language. "What do we mean when we say . . . ?" The "we" is, one would think, ordinary people. The trouble is, with the possible exception of Austin, ordinary language philosophers have never been inclined to do the sort of empirical research necessary to discover how ordinary people use the terms under investigation. Instead they plumb their own usage, assuming that it is "typical." The justification seems to be that if there were no common usage widely spread throughout a language community, people belonging to that community could not communicate with each other. Since they can, there must be some core meaning to every term, if we just work hard enough to uncover it.

The preceding argument should sound familiar. It is the same one used to insist on a core set or cluster of genes and/or traits for biological species. If all organisms belonging to the same species do not share at least a common core set of genes, they could not mate with each other. Similarly, if all members of the same language community do not share at least a common core set of meanings, they could not communicate with each other. Just as this argument is fallacious with respect to mating, it is fallacious with respect to communicating. People speaking the "same" language are broken down into partially overlapping language communities, each speaking its own dialect. Any two English-speaking people chosen at random will agree in what they mean by a vast number of English terms, but it does not follow that there is a common core of terms about which all English-speaking people agree.

But don't ordinary language philosophers, by introspection, discover such core meanings? I don't think so. I think they "construct" such core meanings much more than they "discover" them. As S. Thomason (1991) notes, linguists are never satisfied with hypothetical examples. They insist on real examples as well, because linguistic introspection is so unreliable. For example, according to Chomsky, the English sentence *We received plans to kill me* should be grammatical, while the sentence *We received plans to kill each other* should be ungrammatical (Thomason 1991, p. 255). When tested, Chomsky's students agreed with Chomsky's linguistic intuitions, but when ordinary people were presented with these two sentences, they came to just the opposite conclusion. In the *Meno* did Socrates really get the slave boy to *recollect* geometry, or did he teach the boy geometry through a series of leading questions? From my own experience in graduate school, professors force their students to adopt their linguistic intuitions under the guise of discovering common conceptions (Sorenson 1992, p. 272, terms this process "coaching").

More importantly, what difference does ordinary language and/or conceptions make? If an example shows that a particular scientific theory is internally inconsistent, well and good, but too often the incompatibility is between a technical area of investigation and common sense. As Brown argues, Schrödinger's cat "did not show that the Copenhagen interpretation is logically inconsistent" but rather that it is "in flagrant violation of well-intrenched common sense" (1991, p. 123). However, many of the major advances in science have come at the expense of common sense. Space does not seem to be "curved," but it is. According to ordinary usage, tomatoes are not fruit, as botanists insist, but vegetables. A division of linguistic labor is clearly necessary. Even though we all speak some ordinary language or other, it does not follow that ordinary languages are somehow prior to the technical languages developed for special purposes – and vice versa. Botanists have a reason for grouping tomatoes with fruit. Even so, I do not want tomatoes mixed into my fruit salad. Philosophers have as much right to develop technical languages as do scientists, literary critics, and what have you. I only wish that this school of philosophers had not chosen the title "ordinary language philosophy," when what they are doing is constructing languages that are anything but ordinary.[3]

In response to the claim that species are spatiotemporal particulars, Bradie (1991b) set out a series of science fiction fantasies designed to challenge this conception. His first scenario is designed to challenge contemporary biological understanding. What if spontaneous generation were much more common than we think it is? Organisms sprout up all over the place from inanimate substances. Would evolutionary biologists still insist on a genealogical criterion for species? This scenario strikes me as similar to J. J. Thompson's people-seeds. Suppose "people-seeds drift about in the air like pollen, and if you open your windows, one may drift in and take root in your carpets or upholstery" (1971). Would you be responsible for taking care of such babies or could you mow them down with the vacuum cleaner? In both cases, I think that the answer is the same. The conception of species as individuals depends on the world being basically the way that present-day evolutionary biologists think it is. If natural phenomena were very different from what they are, if natural selection played no role in biological evolution, then all bets are off. Evolutionary biologists have fashioned their species concept to reflect the world as it is, not as it might be. Similarly, our morals depend to a large extent on the world being the way it is. In a very different world, we would have very different morals.

209

Bradie's other scenarios trade on the difference between our ordinary conception of what it is to be "human," assuming that there is such a thing, and someone's belonging to the biological species *Homo sapiens*. Scenario two concerns a mysterious stranger who comes to a small town in the Old West. If this stranger looked and acted pretty much as other people do, then he would be accepted as an ordinary human being. The inhabitants of this small town would have no reason to suspect otherwise. But what if he were a visitor from another planet? What if an extraterrestrial landed here on Earth and mated successfully with human beings, and these hybrids were fertile and mated with other human beings? As strange as it might seem, the answer depends on how common such occurrences turn out to be. Forget about small towns in the Old West and visitors from outer space. Biologists are already aware of cases in which an occasional member of one species mates successfully with members of another species. If such matings become sufficiently common, then the two species merge into one. If they remain rare and sporadic, the species remain distinct.

But what of the particular individuals involved? Wouldn't I be willing to term this stranger "human" if he looked like, sounded like, tested out as, etc., an ordinary human being even if he didn't have the appropriate pedigree (Bradie 1991b, p. 249)? From a variety of ordinary perspectives, possibly so. I know what people are talking about when they say that certain people are "inhuman" and organisms that belong to other species are "human" or "almost human," but from the perspective of evolutionary biology, other considerations are more fundamental. Apparently this stranger was part of one genealogical network. Now he has become part of another – regardless of what he might look like. "But which network does he belong to?" He used to belong to one network; now he belongs to another. As disconcerting as this answer may seem, nothing more can be said on the subject.

Prior to Darwin, biologists found both the splitting and merger of species unpalatable. Darwin forced us to acknowledge that species split, but resistance to merger persists to the present. (Although the splitting and merging of people pose serious problems for the notion of personal identity, evolutionary biologists must consider it because species as lineages both split and merge.) One reason biologists, especially those interested in reconstructing the past, have tried to avoid acknowledging the existence of hybrid species is that they make reconstructing the past by the comparative method impossible. The comparative method is designed to uncover sequential splitting by means of character covariation. Any merger destroys the data necessary to reconstruct the past. Unfortunately, hybrid species do exist today, and we have no reason to expect that they were any rarer in the past (Masterson 1994). As Thomason notes, linguists have resisted the idea of genuinely hybrid languages for the same reason – the problems they pose for the comparative method (1991, p. 251). The arguments that linguists present to prove that hybrid languages are impossible seemed conclusive, but Thomason was able to find a genuinely hybrid language. As unlikely as it may seem, she found a language whose vocabulary came largely from one language and its grammar largely from another.

In his third scenario, Bradie asks what I would say if we gradually replaced the parts of a human being with plastic parts that somehow worked as well as the original parts (1991b, p. 249). The net result looks and acts in every respect like the original. Isn't it still human? Two questions are involved: is it the *same individual* and does it continue to belong to the *same kind*? The first question turns on the old problem of the Ship of Theseus. With respect to ordinary organisms, there is no problem. We remain numerically the same individuals even though we exchange our constituent parts several times over during the course of our existence – just so long as the process is sufficiently gradual and we remain sufficiently cohesive during such change (Wiggins 1967). Organisms can also change their kinds. Certain organisms start off as saprophytic. Later they become parasitic. In this process they remain numerically the same individual. However, "saprophytic" and "parasitic" refer to *ecological kinds*, not *genealogical networks*. Changing one's kind is a different sort of process from moving from one network to another (see Hull 1987).

But what if we made a baby from scratch using plastic parts? It would lack the appropriate genealogical connections, but might it not be considered human? If it matured and mated successfully with other unproblematic human beings, then, yes, it would have become part of the human genealogical nexus. However, there are many problems with these examples for which Bradie gives no hint of an answer. DNA is not a plastic. If the adult in the preceding example eventually comes to be made entirely of plastic as is the baby right from the start, then the genetic material of these individuals cannot be DNA. How fetuses are going to develop when half of their chromosomes are composed of DNA and the other half some sort of plastic, I do not know. But such "details" to one side, any entity that joins in a genealogical nexus is part of that nexus. If species are identified with such chunks of the genealogical nexus, then plastic entities belong in the same species with their organic brethren. "But that don't sound right." That cannot be helped.[4]

Fictitious examples can always be made to fit one's general thesis, while real examples always run the risk of counting against it. For example, Wilkes (1988, p. 15) takes for granted that "human being" and "*Homo sapiens*" denote one and the same natural kind. She argues at considerable length that "person" does not denote a natural kind of any sort (see also Churchland 1982). Even so, she thinks there is "very substantial intersection" between the class of persons and these other two coincident classes (Wilkes 1988, p. 22). This coincidence in turn allows us to "bring to bear the (relative) clarity of a natural-kind-term – *homo sapiens* – on to the puzzling intricacy of the everyday notion 'person'" (ibid., p. 28; see also Churchland 1982). If species were natural kinds, Wilkes's strategy would look promising, but if species are historical entities (space-time worms) and not natural kinds, then her line of reasoning does not begin to get off the ground. Because thought experiments are devised to support a particular position, they are unlikely to challenge it – although sometimes authors are unaware of the implications of their own brain children. Calling for philosophers to make greater use of real examples is a double-edged sword. Real examples are as likely to count against a view as for it.

If thought experiments and science fiction examples are supposed to expand and clarify our conceptual systems, then the reader will have to decide how well these examples served. I for one find general explications of the issues, minus the fictitious examples, much more explanatory.[5] Such examples trade on intuitions, but I am suspicious of intuitions, my own as well as others'. Can't I conceive of species as natural kinds in a significant sense of "natural kind"? Certainly, but what I cannot conceive of is species as natural kinds evolving by natural selection. I cannot conceive of species being both spatiotemporally *restricted* and spatiotemporally *unrestricted*. People who think that they can conceive of this state of affairs do not know enough about the evolutionary process. They think they are conceiving of natural kinds evolving by natural selection, but they are mistaken. All they need to do to discover this mistake for themselves is to learn a little bit about biological evolution.

To the extent that intuitions have any warrant, this warrant resides in the conceptual systems which give rise to them. Given Putnam's theory of reference, the claim that any of us is a brain in a vat is incoherent. "The sense in which I cannot be a brain in a vat is the sense in which it is impossible for me to make a true statement that I am" (Collier 1990, p. 415). But someone might be able to construct a system in which the idea would be unproblematic. Given Chomsky's analysis, certain sentences will seem grammatical, others not. Given the context of the comparative method, hybrids seem impossible. But frequently what is involved is the *replacement* of one conceptual system with another or a major *revision* of an existent system. In such cases, intuitions drawn from one conceptual system are not dependable in evaluating another.[6]

NATURAL KINDS

The preceding discussion concerned scientific concepts – species and *Homo sapiens*. Physicists get to decide what counts as pions, black holes, and charm. Biologists get to decide what counts as cistrons, hybrid sinks, and biological altruism. But "natural kind" is a philosophical concept, and here philosophers are the relevant experts. Philosophers through the centuries have developed the distinction between individuals and classes, primary and secondary substances, etc., and they have done so in the midst of a half dozen or so paradigm examples. Among the stock examples of individuals are particular organisms such as Sea Biscuit and Gargantua. The three most commonly cited examples of kinds are geometric figures, biological species and, a poor third, physical elements.

Scientists decide what is to count as a planet or a biological species in the context of the theories that they develop, and these theories inevitably depart from common sense or ordinary experience. Hence, some examples that sound perfectly cogent in the context of ordinary experience must be dismissed as mistaken regardless of what ordinary people might think. But, conversely, because scientists possess very general, highly articulated theories, they can afford to appeal to fictitious examples with little fear that they will do too much harm, and they might even do some good. The same observation holds for philosophers who, in the good old days, produced full-fledged, well-worked-out philosophical systems. Within the context of these systems, concepts can be explicated by trotting out a string of examples, including hypothetical examples.

But these days are past, at least among analytic philosophers. Global philosophical systems are out of fashion; we limit ourselves to bits and pieces. In the absence of general systems, however, whether scientific or philosophical, how are we to evaluate hypothetical examples? How can we decide what the implications of a particular example actually are? Ordinary language will not do because it is too various, heterogeneous and untechnical. We philosophers have our own technical vocabularies, and we are not about to bow to ordinary usage if it gets in the way of our explications. The distinction between primary and secondary substances, individuals and classes, etc., has been fairly prevalent in philosophy throughout its history. Of course, no two philosophers meant the same thing by "kind" or "natural kind" and their cognates in other languages. Is there an essence of "natural kind" as philosophers have used this phrase through the years? Quite obviously not. In the absence of such universal usage, I am forced to set out what I take to be a fairly prevalent usage to see the effect that hypothetical examples have on this conception.

Natural kinds has been commonly understood to have three defining characteristics. They are eternal, immutable and discrete. In what sense are geometric figures, biological species, and the physical elements eternal, immutable and discrete? By "eternal" some philosophers meant that every natural kind is always exemplified. Horses always were and always will be. However, many philosophers have had a weaker notion in mind. Natural kinds are built into the structure of the universe. At times they may be exemplified; at times not. During the first few minutes after the big bang, none of the heavier elements existed. If sexual reproduction involving reduction division is necessary for biological species, then for most of life on Earth, there were no biological species. Meiosis is a relatively recent innovation. Natural kinds can be "eternal" in this sense even though they periodically are not exemplified. Gold can come and go.

Some philosophers have held a very strong notion of "immutability." All individuals belong at any one level to one and only one natural kind, and no individual can change its kind. Water snakes never become vipers, nor do frogs ever become toads. Other philosophers are more liberal. A sample of lead might be transmuted into a sample of gold. Generations of alchemists tried. As it turns out, such transmutations are possible. The main drawback is that producing gold in this way is much more expensive than mining it. Similarly, a wire circle can be reshaped into a square. But in both of these cases, individuals are changing their kinds (although philosophers disagree about whether or not they remain numerically the same individuals). The kinds themselves remain unchanged. Lead qua lead is not evolving into gold qua gold, nor are we squaring the circle.

Finally, philosophers have frequently claimed that natural kinds in conceptual space are discrete; that is, each natural kind can be defined by a set of necessary and sufficient conditions. The entities that exemplify these kinds, however, need not do so perfectly. Some deviation is only to be expected. No triangle is perfectly triangular, no sample of gold is totally pure, and no horse perfectly exemplifies the essence of horse. The assumption is, however, that borderline cases are relatively rare. If mapped in conceptual space, most entities cluster around their norm, and each kind has only one norm. In a very real sense, any *variation* is necessarily *deviation*. A horse lacking a tail is a deviant horse, women are deviant males, blacks are deviant whites, and homosexuals are deviant heterosexuals. In this context, neutral references to variation are inappropriate.

The assumption in the preceding discussion is that geometric figures, biological species, and the physical elements are equally natural kinds, but if we look at these three examples with care, we can see that they form a heterogeneous hodgepodge. If the distinction between an uninterpreted formal system and a physical application of that system is taken seriously, then Euclidean geometry is very different from applied Euclidean geometry. Euclidean parallel lines never cross, but light rays always cross no matter which way they are directed. Euclidean triangles and triangles made of light rays are very different entities. They may be natural kinds, but they are natural kinds in two different systems. Physical elements do seem to be natural kinds, albeit derivative of the more fundamental natural kinds at the subatomic level. Given the subatomic particles, the physical elements are built into the basic structure of the universe.

But biological species as the things that evolve exhibit *none* of the characteristics of natural kinds. They are not eternal. They come and go, and once gone, can never return. A creature that looks like a dodo, exhibits the same array of genotypes as the dodo, etc., may come into existence. But, so we are told, the dodo is extinct, and "extinct" does not mean temporarily not exemplified. Conservationists are right to worry about endangered species; like herpes (and unlike true love), extinction really is forever. The dodo can no more re-evolve than Hitler can come into existence again. Nor are biological species in any sense immutable. One way of going extinct is to evolve into two or more descendant species. Finally, they are not discrete. To the extent that evolution is gradual, species when mapped onto conceptual space have numerous intermediates. In physical space, most species are reasonably discrete. Their ranges have fairly sharp limits, albeit limits that can change with the seasons. The relevant "gradualness" concerns one species changing into another species, either through anagenetic or cladogenetic modification. Even if speciation is usually punctuational, intermediates still exist because the mechanism set out by Eldredge and Gould (1972) is populational. Some species, however, do come into existence saltationally. For example, in cases of allopolyploidy, a new species can come into existence in the space of a single generation.

At this juncture, several alternatives are possible. Some authors, such as Dupré (1993), retain traditional conceptions of natural kinds and use the failure of biological species to exemplify this conception as one more argument against the existence of natural kinds. Another alternative is to loosen the notion of natural kind to include species. The trouble is that the degree of loosening necessary to include species results in organisms and just about everything else becoming kinds. The alternative I prefer is to leave the notion of natural kind as it is and admit that we got one of our examples wrong. On this alternative, biological species no longer pose any problem for the notion of a natural kind because they are not natural kinds.

One reason an adequate definition of "natural kind" seems so difficult to come by is that we have tried to formulate this definition on the basis of a motley collection of examples. Perhaps ordinary people think of geometric figures, physical elements and biological species as the same kind of thing, but ordinary conceptions are not good enough for either science or philosophy. At times they are not good enough for everyday life either. Just as philosophers have tried to distinguish between genuine laws of nature and accidentally true universal generalizations, we have also tried to explicate the parallel distinction between natural kinds and artificial collections. In both cases, there is a felt difference, but can we set out any criteria to justify these feelings?

One way to mark the latter distinction in science is to argue that natural kinds are kinds that function in natural regularities. (I am not *endorsing* these distinctions but pointing out how they might possibly be salvaged.) A kind term is "natural" if it occurs in the statement of a natural law. On this analysis, the statements of uninterpreted calculi do not count as laws of nature. Hence, geometric figures do not form natural kinds. Some other justification is needed for them. Because biological species are not kinds at all, they cannot function as natural kinds. Statements such as "All swans are white," even if true, would not count as natural laws. Lawful regularities in the living world occur at higher levels of analysis. Although the physical elements are not especially fundamental kinds, they are nevertheless natural kinds on this analysis because the names of natural kinds function in natural laws.

Distinguishing between natural kinds and accidental collections via natural laws points up, quite obviously, the need for an adequate analysis of natural laws. (Once again I am not endorsing the legitimacy of either natural kinds or laws of nature but describing how they might be related if one accepts them.) None of the usual criteria that have been suggested works very well. The best is that natural laws are the generalizations that function in scientific theories. Any generalization that remains isolated, as true as it may be, is likely to be an accidentally true generalization. Even if this criterion turns out to be adequate, the next move is, of course, to demand an analysis of scientific theory. One way is to explicate the more inclusive notions in terms of their constituents. Scientific theories are made up of scientific laws, and laws are those generalizations that refer to natural kinds, never accidental collections. As always, the two ultimate alternatives are infinite regresses and circular reasoning. Outside these two alternatives reside pragmatic justifications – how well these analyses accommodate examples, and if this essay has any message, it is that these examples have to be real. They have to be part of some well-articulated system, whether scientific or philosophical.

How about fictitious examples? Don't they work just as well? In a celebrated paper, N. Goodman (1954) presented a contrary-to-fact example to pose a new riddle of induction. Such color predicates as green and blue seem like perfectly natural ways to divide up the world. Goodman suggests other predicates that seem decidedly unnatural – grue and bleen. "Green" applies to everything that is green – past, present and future, whether examined or not, and the same for "blue." "Grue," however, applies to things examined before a particular time, say, 2000 A.D., just in case they are green but to things not examined before 2000 A.D. just in case they are blue. Thus, anything examined before the year 2000 A.D. is both green and grue. What justification do we have for preferring talk of green emeralds to grue emeralds? That "emerald" is currently *defined* as green beryl is unlikely to give anyone pause.

Dozens, possibly hundreds, of papers have been written about grue and bleen without making much headway. My response will no doubt seem totally wrong-headed.

Genuine change does occur in nature. For example, in the early millennia of life on Earth, respiration was entirely anaerobic. Somewhere along the way, aerobic forms of life gradually replaced anaerobic forms. In addition, life on Earth has been punctuated by mass extinctions in which upwards of 95 percent of all species went extinct. Given our current understanding, such occurrences are not anomalies. They are the sorts of thing that can happen, even though scientists are currently arguing about the mechanisms that actually produced them. However, according to current physical theory, predicates do not and cannot act the way that grue is supposed to act. There is nothing special about the year 2000 A.D. If things change their color wholesale on New Year's Eve of the second millennium (or, more accurately, if we change how we conceive of colors), then we have seriously misunderstood nature and have to reevaluate our beliefs from the ground up. Since our intuitions are informed by what we believe, in this new world we can no longer depend on our intuitions, but this is precisely what Goodman asks us to do.

"But can't you conceive of an alternative theory in which the year 2000 A.D. is as natural as absolute zero?" Once again, I have to say that I cannot. I can conceive of several generations of physicists producing such a theory but not me off the top of my head – and that is what appeals to such examples require.[7]

CONCLUSION

In this essay I have urged the advantages of real examples over fictitious examples. Both real and fictitious examples can be presented briefly, both can challenge our conceptions, but if the occasion demands, we can expand upon real examples because they appear in a larger context that supports such investigations. Made-up examples that occur outside of any context, including a well-formulated fictional world, can be modified any way one pleases. The considerations that I have sketched in this essay have forced philosophers and scientists alike to reevaluate some fundamental beliefs of Western thought from the ancient Greeks to the present. These considerations concerned the nature of species if species are supposed to be the things that evolve. Perhaps fictitious examples could also have motivated these changes, but I doubt it.

In this essay I have presented both a negative thesis and a positive thesis. The negative thesis is that made-up examples, especially of the sillier sort, have done massive damage to philosophy, in particular the philosophy of science. The chief deficiency of hypothetical examples in philosophy is that no one has bothered to set out a list of criteria by which they can be evaluated. What makes for a good hypothetical example? Why hasn't the Ship of Theseus done as much damage as grue and bleen?

The positive thesis is that philosophers should use real examples whenever they can because they have all of the virtues of made-up examples and none of the vices. While we are at it, we might set out a list of criteria for the evaluation of real examples as well. Another reason for my urging real examples on philosophers is that real examples, unlike fictitious examples, can be used as evidence – but that is another story.

NOTES

Reprinted from *The Cosmos of Science: Essays of Exploration*, ed. John Earman and John D. Norton.© 1997 by the University of Pittsburgh Press. Reprinted by permission of the University of Pittsburgh Press.

I owe a note of appreciation to several friends and colleagues who encouraged me in the writing of this essay: Martin Bunzl, John Collier, Arthur Fine, Tom Ryckman, and Ken Waters. However, I periodically got the feeling that they were encouraging me to be the first penguin off the ice floe.

1. As Norton sees it, the problem is that case studies tend to come in two polar types. "Either they are contrived 'toy' models, whose logical relations are clear but whose connection to real science is dubious. Or they are instances of real science of such complexity that one must be disheartened by the task of mastering the scientific technicalities let alone disentangling its logical structure" (1993, p. 412). One characteristic which real experiments possess and thought experiments lack that I have not discussed in this paper is historicity. Hacking (1993) argues that real experiments "have a life of their own" in a way that made-up examples do not (see also MacIntyre 1981 and Sorenson 1992). However, some made-up examples have been discussed so frequently, later

commentators modifying them for their own purposes, that they have developed a life of sorts.

2. Of course, with a sufficiently broad notion of set (or class, or kind) everything counts as a set from bare particulars to Richard Nixon. My response to this maneuver is to distinguish between two sorts of sets – those that are spatiotemporally restricted and those that are not (see Hull 1978a, p. 340). Species are instances of the first sort of "set." According to current physical theory, the universe is spatially finite and probably temporally finite as well. If so, then all classes are "spatiotemporally" restricted in this sense (Hull 1981a, p. 148).

3. In response to Sober's (1984) analysis of selection in evolutionary biology, Waters objects that Sober "bases his own acceptance of the harsh Pareto-style requirement on intuitions about phenomena that seem to have little relevance to evolution by means of spatially dependent selection. Rather than wage a battle of philosophical intuitions about phones connected to Rube Goldberg devices and baseball fans jumping up in stadiums, I am basing my case against applying the Pareto-style requirement *to selection theories* on the claim that we must abandon this principle to maintain scientific realism. If this means we must compromise some intuitions about population-level causation, then so be it. Philosophers of science have not held intuitions sacred in trying to preserve realism in the face of modern physics. I see no reason why intuitions about nonevolutionary phenomena should be considered gospel when it comes to preserving realism about the force of natural selection" (1991, p. 570).

4. One reason philosophers may be so reticent to treat *Homo sapiens* as a historical entity is that it threatens a central notion of certain ethical systems – human nature. If human beings have no nature, no essence, then all is lost. But just because biological species, including the human species, have no essence, it does not follow that persons, rational agents, or bearers of immortal souls have no essence. The problem is the identification of these kinds with something that is not a kind – *Homo sapiens* (see also Hull 1988a).

5. I am afraid that the acclaim accorded J. J. Thompson's 1971 article stems more from her arresting examples than from her arguments. I find myself in large agreement with her conclusions in spite of her examples, not because of them. Nor does Thompson expect her readers to take her examples realistically, as if they really were to wake up connected to a famous violinist. The typical American reaction would be to get a lawyer and sue the Society of Music Lovers. A few million dollars would be ample compensation for a few months of inconvenience. But Thompson would no doubt dismiss this response, as realistic as it might be, as inappropriate.

6. In two of the examples that I discuss, the conceptual tensions one feels can be reduced to contradictions. Species cannot be both spatiotemporally restricted and spatiotemporally unrestricted. One cannot simultaneously accept a particular set of criteria for reference and accept a term that lacks these criteria as actually referring. Norton (1993) thinks that all reliable thought experiments in science involve nothing but inference. Bunzl goes even

further, arguing that all reliable thought experiments turn on straightfor-
wardly deductive considerations.

7. Akeroyd (1991) has presented a realistic example of a grue-like predicate.
According to Akeroyd, "grue" is defined as green in the past but blue in the
future. He suggests "regulatic" phenomena as more practical examples: those
that "exhibit regularities before time t and erratic behaviour thereafter"
(Akeroyd 1991, p. 535). He then proceeds to give a real example of such regulatic phenomena from economics. Prior to 1969, the relationship between the
percent rate of unemployment and the percentage change of money wage
rates exhibited regular behavior; thereafter, it did not. The economic system
had evolved into quite a different system. In certain economic systems stagfla-
tion is impossible. In others it is perfectly possible.

11

Studying the Study of Science Scientifically

Testing the claims that scientists make is extremely difficult. Testing the claims that philosophers of science make about science is even more difficult, difficult but not impossible. I will discuss three efforts at testing the sorts of claims that philosophers of science make about science: the influence of scientists' age on the alacrity with which they accept new views, the effect of birth order on the sorts of contributions that scientists make, and the role of novel predictions in the acceptance of new scientific views. Without attempting to test philosophical claims, it is difficult to know what they mean.

In the summer of 1996, David Bloor (1997, p. 498) was dismayed to hear a philosopher argue before an international audience that, according to Kuhn, "communication between scientists in different, incommensurable paradigms is impossible, but such communication is indeed possible – and here examples were provided – so Kuhn is wrong." In response, Bloor (1997, p. 499) argues that Kuhn was "not equating incommensurability with incommunicability, though he was saying that it can attenuate communication and may demand an effort to overcome." I do not know whether the philosopher in question was as obtuse as Bloor claims, but he or she was doing something right – attempting to test claims about science made by students of science. If Kuhn is right about paradigms, one would expect that scientists working in the same paradigm should be able to communicate with each other more easily than scientists working in different paradigms. This claim can actually be tested, and the results of this test might reflect positively or negatively on Kuhn's famous incommensurability theses – and without equating incommensurability with incommunicability.

In this paper I urge more students of science, especially philosophers of science, to test their claims about science as often and as rigorously as possible. To perform this task, science must be distinguished from the study of science. History of science poses one set of problems. In certain respects it is like all other historical disciplines, such as cosmology, but human history provides all sorts of other problems in addition. Historians commonly claim that human histories are as much literature as they are science. Sociologists of science are put in the position of having to study themselves. After all, sociology of science is part of sociology, and sociology is itself a science. But of all the disciplines that make up science studies, philosophy of science is furthest removed from science itself. In this paper I concentrate on the hardest case of all – philosophy of science.

Those of us engaged in science studies study science, but what sort of thing is science? In the past, philosophers of science have taken science to be something like a natural kind, about which general knowledge is possible. On this view science is a certain sort of activity distinguishable from other human activities by characteristics such as serious attempts at falsification. At the other extreme, current advocates of pluralism claim that science is merely a "loose and heterogeneous collection of more or less successful investigative practices" (Dupré 1990, p. 69). In this paper I treat science neither as a natural kind nor as a loose heterogeneous collection, but as an "historical entity," an entity that is variable both through time and at any one time but which retains sufficient cohesiveness and continuity in the face of all this variability.

After numerous false starts, science dribbled into existence and, once established, continues to the present. To be sure, it cannot be characterized in terms of universally unvarying processes, but it exhibits much more coherence and continuity than Dupré would allow. At any one time, science is variable. For example, some cognitive psychologists allow reference to introspection while behavioral psychologists do not. Science also varies through time. While reference to God was acceptable in the early years of science, it is not acceptable now. Furthermore, science has not yet reached its definitive state today. It is very likely to continue changing in the future, possibly in ways that few of us would like. For example, political correctness may supersede empirical truth as a criterion for scientific acceptance. If so, my interest in science will decrease markedly. But, as variable as science has been at any one time as well as through time, it is not a hodgepodge. It exhibits enough coherence and continuity to be considered an historical entity.

Regardless of the way in which one construes science, it still can be studied scientifically. If science is a kind, possibly a natural kind, then one task of those of us who study science might well be to discern the regularities that characterize it. If science is an historical entity, we can still study its variability at any one time and its modifications through time. Finally, if science is just a variable and heterogeneous collection of investigative practices, then all we can do is to chronicle its vicissitudes. Thus, on any of these interpretations, science can be studied scientifically because scientific understanding encompasses both the specification of lawful regularities and the description of particular sequences of events. However, if we are to study science scientifically, we must meet the same standards as those that govern science itself. As Sulloway and others have remarked, too often those of us who study science tend to employ methodological practices that we would condemn as inadequate in the work of those scientists whom we study.

In recent years, science studies has been characterized by a welter of conflicting schools, philosophies, and methodologies. According to many students of science, much of what we commonly take to be important in science consists in little more than public relations ploys. Scientists do not discover facts; they construct, fabricate, and invent them. To the extent that these expressions were devised to emphasize the active role that scientists, both individually and in groups, play in the generation of knowledge claims, they are unproblematic; but I personally can understand why many readers, including scientists, take these claims to imply much more, as if nothing that can be legitimately termed "nature" or "the world" in any way constrains what scientists come to believe. If any of the advocates of these philosophies actually hold any of these more extreme positions, they will surely reject my call for students of science to test their beliefs about science as one more instance of self-delusion. If all the time that scientists expend testing their views is just so much show, then surely any attempt by those of us who study science to test our views about science scientifically is even more of an illusion.

The relation between science and philosophy of science is usually characterized in terms of "levels." Science is related to philosophy of science in the same way that ethics is related to metaethics. Philosophy of science is one level "higher" than science. The relation between these levels is problematic. For example, logical empiricist philosophers of science insist that empirical data play a necessary role in science.

However, they need not on this account be committed to the view that empirical data play any role whatsoever in philosophy of science. The Covering-Law Model of Scientific Explanation might be a totally adequate analysis of "scientific explanation" even if no scientist ever explained any phenomenon by deriving it from a law of nature. The central role of evidence *in* science entails nothing about the importance of evidence in our study *of* science. However, this difference between science and the study of science does, to say the very least, call for an explanation. In this paper I see to what extent the sort of testing that goes on in science can profitably be extended to the study of science. Traditionally, historians and sociologists of science have acknowledged the role of data in their work. Philosophers should do the same. Not only should we use the data gathered by historians and sociologists of science, but we should also generate a bit of this data ourselves. In doing so, we will deepen our understanding of both science and philosophy of science.

One message that both history and sociology of science have to teach us is that testing is difficult. These difficulties have several sources. One difficulty arises from the intricate interconnections between our basic theories and anything that might count as empirical data. It is simply not true that we know what our theories mean independently of any attempt to test them. We may think that we understand a particular theory, but this illusion can be quickly dispelled by attempting to test it. No amount of conceptual analysis can serve as a substitute for empirical investigation – even with respect to meaning. Empirical investigation requires the operationalizing of the concepts we are attempting to apply. Philosophers have contributed to our understanding of science by showing that theoretical terms cannot be operationally defined in a literal sense of "definition," but this impossibility proof is not enough. We still need some discussion of how to operationalize our concepts. What difficulties are we likely to meet? Can they be overcome?

Testing the sorts of claims that scientists make is difficult enough. Testing the sorts of meta-level claims made by those of us who study science is even more difficult. Even so, we must try. In this paper I investigate three attempts to study science scientifically: my own study of the role of age in determining how quickly scientists accept new ideas (Hull, Tessner, and Diamond 1978), Sulloway's (1996) attempt to discover the influence of birth order on the sorts of contributions that scientists are likely to make to science, and Donovan, Laudan, and

Laudan's (1988) investigation of the role of novel predictions in science. By discussing these three examples, I intend to show how general claims about science can be tested and to illustrate how the sorts of difficulties that arise in such attempts can be overcome. They are overcome in the same ways as they are in science in general. By emphasizing the role of testing via data, I do not mean to imply that this is all there is to science. Obviously, science is much more than empirical testing. However, such testing is an important aspect of science. Also, in emphasizing the need for philosophers of science to attempt to test their views about science, I do not mean to imply that this is all that philosophers of science should do. Occasional testing does not entail nothing but testing.

PLANCK'S PRINCIPLE

In his autobiography, Max Planck (1949, p. 33) remarked that a "new scientific truth does not triumph by convincing its opponents and making them see the light, but rather because its opponents eventually die and a new generation grows up that is familiar with it." Instance after instance can be given of scientists, who, like Planck, were involved in a basic conceptual revolution in science, claiming that older scientists are less willing to adopt new views than their younger colleagues.[1] This claim is important because it casts doubt on the role of reason, argument, and evidence in bringing about change in science. Good evidence should be just as good for young scientists as for their older colleagues. If older scientists are more resistant to new ideas than are younger scientists, it might be due to the biological or psychological effects of aging, to differences in the stage of a scientist's career, to professional commitments to certain views, or to greater knowledge of the effects of this change, and so on (for discussion of the role of age in science, see McCann 1978; Messeri 1988; Rappa and Debackere 1993; and Levin, Stephan, and Walker 1995).

The investigation of Planck's principle raises two questions: what does it mean? and is it true? Until we begin to decide what Planck means, we cannot begin to decide whether or not Planck was right. What counts as a "new scientific truth"? How "new" must it be? To answer these questions, I turn to two examples of scientific revolutions: the Darwinian revolution and the Mendelian revolution. Inevitably, when scientists publish what they take to be new scientific ideas, historians will find all sorts of precursors. For example, Matthew (1831) anticipated Darwin with respect to natural selection, and numerous authors, including Lamarck and Robert Chambers, anticipated him with respect to evolution. Why then term the revolution that took off in 1859 the *Darwinian* revolution? Why not the Matthewian revolution? The answer is that Matthew did not produce a revolution of any kind. His allusions to what came to be known later as "natural selection" went totally unnoticed at the time. Authors such as Lamarck and Chambers had some impact with respect to the transmutation of species (e.g., on Wallace), but neither succeeded in producing anything like a "revolution." Darwin did.

The Mendelian revolution poses a different set of problems. Was Mendel the author of the Mendelian revolution or was he simply a precursor like Matthew, Lamarck, and Chambers? Although Mendel's 1865 paper did not go totally unnoticed, it certainly did not initiate anything like a revolution in the study of heredity. The Mendelian revolution did not occur until the turn of the century when others came up with views similar (though far from identical) to those of Mendel. These authors claimed independent discovery. If so, then Mendel's paper played no role in the "rediscovery" of Mendelian genetics. Then why call this revolution "Mendelian"? One answer is that Mendel's paper includes a careful and clear exposition of the theory that eventually came to bear his name. It was not just a vague sketch. Another reason is that it served to stave off a priority dispute. The two junior rediscoverers were not about to have this theory named after them. They were able to sabotage any potential effort by the powerful de Vries to get his name attached to this emerging field by emphasizing the role of Mendel. Better Mendelian genetics than de Vriesian genetics. The sense of novelty that is relevant to Planck's principle concerns *reception*, not first discovery or even first publication. Until scientists notice a new idea, they cannot accept or reject it.

What counts as a *scientific* truth, and must it be *true?* Theologians, art critics, wine connoisseurs, and hosts of others may or may not be resistant to change, but Planck's principle applies only to scientists. Who are we to count as scientists? In evaluating Planck's principle, are we to count scientists as they were conceived of at the time or as we conceive of them today? After all, who is or is not considered a scientist has changed during the course of science. For example, Descartes, and later Newton, were engaged in roughly the same array of activities. They were engaged in what we today would call philosophy, mathematics, and science, not to mention theology, alchemy, and other questionable practices, but the preceding distinctions were not made at the time the way that we make them today. Nowadays, college students are likely to come across Descartes in an introductory philosophy course, while they hear about Newton as a scientist. Why is this so? The reason is obvious: Descartes generated what we would now call a philosophical research program (only specialists pay any attention to Descartes's science), while Newton generated a successful scientific research program (again, only specialists are much interested in Newton's philosophy). Hence, today, Descartes is usually characterized as a philosopher and Newton as a scientist. But such presentist ascriptions are sorely misleading.

I see no easy way out of the dilemma posed by the evolution of language. All of us are saddled with the requirement that we write for present-day readers and must use present-day language to do so, even if we know that present-day language is sure to mislead. The best we can do is to warn our readers of the most obvious and important instances of possible misunderstandings. After all, publishing a book on sixteenth century Italian science in sixteenth century Italian would be a waste of time. Those who could make the most sense of such a book have long been dead. Even the most anti-presentist historians now see the need for what they term "legitimate anachronisms" (Lightman 1997, p. 10). However, we must constantly remind ourselves that calling Descartes a philosopher and Newton a scientist in the absence of such explanations can lead to serious misunderstandings.

With respect to Planck's principle and the reception of Darwin's theory of evolution, we cast our net broadly. We included anyone who was treated as a scientist between 1859 and 1869 by their contemporaries, even if we had our doubts. Happily, we may term these workers "scientists" because William Whewell first coined this term in 1833 and published it in 1834, only to reject it as inappropriate (Whewell 1834). It caught on anyway but not until a half century later. Why we are permitted to use a term only after it was introduced – even if rejected by its author and by the relevant audience for half a century – remains, for me, one more historiographic mystery. In any case, we were well aware that we were not sampling the scientific community at large. Those scientists who were most likely to pass judgment on evolutionary theory were those most closely connected to it conceptually, e.g., zoologists, botanists, geologists, those physicists working on the age of the earth, etc.

In addition, well-known scientists are likely to leave more extensive records than their lesser-known brethren. This second limitation might seem on the surface not to be of great significance, except for the work of Desmond (1989) and Kim (1994) with respect to the reception of Darwinian evolution and Mendelian genetics, respectively. They discovered that second- and third-tier scientists reacted quite differently from their more prominent colleagues. Although Darwin valued the opinion of Charles Lyell over that of an obscure pigeon fancier, Planck clearly intended to be talking about scientists, not just the scientific elite (for discussions of the elitist nature of science, see Meltzer 1956; Price 1963; Cole and Cole 1973; Yoels 1973; Reskin 1977; Ladd and Lipset 1977; Garvey 1979; and Hull 1988a).

Are older scientists resistant only to new *truths?* Planck's principle is usually raised only with respect to new scientific ideas that we now take to be true or at least significant improvements over past beliefs. How about all the new ideas that are rejected, and, in retrospect, we think rightly so? Rarely does anyone complain about the resistance of scientists to phrenology, the Piltdown man, polywater, cold fusion, and a host of other ideas that we now take to be mistaken. Instead, the usual question is why serious scientists were taken in. Are older scientists more able than younger scientists to tell truth from falsity? Or are older scientists supposedly more resistant to all new ideas, independent of the eventual fates of those ideas? On my reading, Planck's principle concerns the latter. Once again, older scientists as they were conceived of at the time are supposedly more resistant than younger scientists to *all* new ideas, regardless of how we judge, in retrospect, the truth-values of these ideas.

Idea and theories are very difficult to individuate. How much of a new theory must other scientists accept in order for us to count them as accepting that theory? Darwinism, for example, is a very general, heterogeneous theory. Different elements have had very different fates. If one of Darwin's contemporaries had to agree with Darwin on all his basic tenets in order to be counted as accepting Darwin's theory, then very few scientists could be considered Darwinians. (Numerous evolutionary biologists can be found making claims about what the "essence" of Darwinism is, but such pronouncements are simply efforts to throw Darwin's mantle over their own shoulders.) In order to test Planck's principle for something like Darwinism, one element of the general theory must be singled out, for example, the basic claim that species evolve. One might disagree with Darwin about the role of sexual selection in evolution and still count as accepting his theory, but it would be difficult to consider anyone a Darwinian who refused to acknowledge that species evolve. Of course, Darwin's belief that species evolve was the least original part of his theory (for further discussion, see Hull 1985a and Numbers 1998).

Must older scientists literally *die?* Retirement from scientific life is good enough. The relevant generation time is professional, not biological. After all, roughly 75 percent of professional scientists had accepted the evolution of species by 1869 without any upsurge in the death rates of scientists (Hull, Tessner, and Diamond 1978). In different areas of science, periods of productivity vary tremendously. A scientist who is thirty years old might be considered young in one field, over the hill in another (see Merton and Zuckermann 1973; McCann 1978; Messeri 1988; and Rappa and Debackere 1993 for further discussion).

The preceding discussion might sound as if it were designed to discourage anyone from attempting to test hypotheses about science, but it is not. In spite of the need to "operationalize" hypotheses about science in order to test them, testing can still be done. Scientists do it repeatedly in their various areas of science. Students of science can do the same. After all, two decades ago, Tessner, Diamond, and I were able to reach at least some tentative conclusions about the applicability of Planck's principle to the Darwinian revolution, and the sophistication of the literature has only increased since then. For example, we discovered that when the mean age in 1859 of scientists who accepted evolution between 1859 and 1869 is compared to that of those who still held out after 1869, the results are statistically significant (39.6 to 48.1). However, if one looks just at scientists who came to accept evolution between 1859 and 1869, age accounts for less than 10 percent of the variance. Of course, in studies such as these, 10 percent is substantial. It is also true, as pointed out by Sulloway, that small variance does not imply small effect size.

However, the correlation between age and acceptance of the evolution of species after 1859 is not as dramatic as participants and later commentators alike have claimed. It is statistically significant but not so overwhelming that participants should have picked it up through casual observation. My suspicion is that revolutionaries such as Planck and Darwin were not reacting to the resistance of older scientists as such but to the resistance of the leaders in their respective fields – the scientists whose ideas were under attack. The relevant issue is professional age and standing in the community, not biological age. Elite scientists reacted negatively to the views of Planck and Darwin, not because they were old but because they were the authors of the views that were under attack. For the same reason, prestigious professional journals tended to ignore Darwin's theory (Burkhardt 1974).

The message of the preceding example is that even a claim as apparently straightforward as Planck's principle requires extensive reformulation and operationalization before it can be tested. As a result, any attempts to test hypotheses about science are likely to result in their being modified extensively. In attempting to test Planck's principle, I was forced to address questions that would have never occurred to me otherwise. I understand the principle much more deeply and thoroughly after testing it than before. I now have good reason to think that, on certain interpretations, Planck's principle is true; on other interpretations, it is false. The important consideration involves differences in interpretations. Although we all thought that we understood Planck's principle, we were seriously mistaken. Vague ideas floated before our eyes, and that is all. (For more recent discussions of Planck's principle, see Rappa and Debackere 1993; as well as Levin, Stephan, and Walker 1995; and Diamond 1997.)

BIRTH ORDER AND SCIENCE

The issue that I investigated in the preceding section is as narrow and tractable as any issue is likely to get in the study of science. Others are sure to be much more complicated. Almost three decades ago, Sulloway began to investigate the effects of birth order on science. In the course of his investigations, he has expanded his study to include over forty variables, twenty-eight scientific controversies, and hundreds of scientists (Sulloway 1996). Starting with evolutionary theory, Sulloway discovered a marked tendency for laterborns to accept Darwin's theory, but as he expanded his investigation to include other scientific controversies, a more complicated correlation emerged. New theories with a strongly conservative cast attract firstborns, while those that are ideologically radical attract laterborns.

At first glance, deciding birth order would seem to be as straightforward a task as determining a scientist's age, but Sulloway soon discovered that it is much more complicated than one might expect. For example, the sex of one's siblings, the spacing between siblings, one's precise birth rank, and overall number of siblings all turned out to matter. In male-dominated societies, having an older brother is quite different from having an older sister. In addition, Sulloway discovered that a firstborn whose closest sibling is more than six years younger functions as if he were an only child, whereas twelve years have to elapse before a laterborn functions as an only child. Although Louis Agassiz was his parents' fifth child, he functioned as a firstborn because the other four died in infancy. Parents also remarry, combine and separate families, and much, much more. Sulloway had to make decisions on these and other issues, including some of the same issues that I had to address with respect to Planck's principle. Who is to count as a scientist or as accepting Darwin's theory? On the second score, he opted for more stringent requirements than I did. In addition to accepting the evolution of species, a Darwinian had to acknowledge natural selection as an important factor in evolution (Sulloway 1996, p. 29).

More problematic still is rating the controversies on an ideological scale running from conservative to radical. Sulloway opted to have experts make these decisions intuitively based on their own expertise. Although lots of in-principle arguments can be raised against the use of expert opinions in making such a rating, the results are more important than the arguments. In Sulloway's study, the average interrater reliability turned out to be quite high. At the very least, these authorities shared common biases. As impressionistic as expert opinion may be, it has proved to be reliable in certain contexts. For example, practicing taxonomists are very good at estimating taxonomic relationships – except in very special circumstances – and those systematists who have attempted to quantify taxonomic judgments have gradually uncovered what these special circumstances are (see, e.g., Fernholm, Brener, and Jörnvall 1989).

The most important bias that present-day observers are likely to introduce stems from our presentist perspective. Although Sulloway acknowledges such difficulties, I think they pose one of the most serious problems for his undertaking. He treats certain themata as if they were independent of time and place. Firstborns tend to prefer continuity, order, causality, hierarchy, and essentialism, while laterborns are inclined toward discontinuity, chaos, acausality, equality, and population thinking. However, the assertive content of these themata seem to have changed through time. For example, postmodernist radicals take "essentialism" to be the primary source of the evils that have afflicted societies throughout history. Hence, they would be counted as confirming Sulloway's hypothesis. But as postmodernists use "essentialism," it does not contrast with population thinking but with realism, and realism is not on Sulloway's list. Idealism has raised its head periodically in the course of science. In Darwin's day it was a conservative view, but today it is quite radical (Webster and Goodwin 1996). How are we to score scientists from one age to the next?

I do not raise these problems in order to discount or reject Sulloway's studies. To the contrary, these issues arise only because Sulloway chose to take birth order seriously. Nor do I think that these problems are insuperable. The reason that certain students of science are likely to resist studies such as those conducted by Sulloway is that they seem to imply that certain factors, which we traditionally think do not or should not have a significant effect on science, actually do. However, the traditional view of science is not in the least threatened if increased financial support for science leads to its acceleration, or if firstborns tend to be overrepresented in the rank of eminent scientists. Truth is truth regardless of the birth order of those who discovered it. However, the correlations that Sulloway found between birth order and preference for certain sorts of ideas do pose problems. If firstborns tend to have disproportionate effects on science and prefer certain ideas, then the cognitive content of science is likely to reflect these preferences. If science is to reflect the world in which we live more strongly than the character of scientists themselves, it must incorporate one or more mechanisms capable of neutralizing the effects that Sulloway has discovered (see Ruse 1996 for an exhaustive detailing of the effects that a belief in progress has had on science, especially evolutionary theory).

THE ROLE OF NOVEL PREDICTIONS IN SCIENCE

Over a decade ago, the people at Virginia Polytechnic Institute and State University decided that it was about time to subject some of the key claims made by theorists of scientific change to the same kind of "empirical scrutiny that has been so characteristic of science itself" (Donovan, Laudan, and Laudan 1988, p. 3). In the past, students of science have sporadically attempted not just to illustrate general principles about science but also to test them by reference to episodes in science. The trouble is that the results of these studies are not comparable. They embody different assumptions, attempt to test different principles, and proceed in very different ways. In this respect, science studies is not all that different from other areas of science. Certain naive investigators (e.g., Collins 1985) were shocked to discover that scientists do not replicate each other's work all that often or all that precisely. They tend to concentrate on results that conflict with their own views, and when they do run these experiments, they are rarely narrow replications. Scientists modify certain aspects of the experimental setup for a variety of reasons. However, in many cases, these differences are so slight that the studies can be compared. One problem with the empirical studies that have been done in the study of science is that they are so different that they cannot be meaningfully compared.

The folks at Virginia Tech decided to remedy this situation. They gave over thirty historians, philosophers, and sociologists of science the task of evaluating a variety of claims commonly made about science by members of the post-positivist historical school in the philosophy of science. They ended up publishing the results of only sixteen of these case studies (Donovan, Laudan, and Laudan 1988). The organizers of this science studies research program were well aware that their activity was necessarily reflexive. For example, one of the theses of the historical school is that no thesis can be expressed in an entirely theory-free vocabulary. Popperians use one vocabulary, Kuhnians another, Laudanians yet another, and so on. If the claims to be tested are set out in Popperian terms, then the Popperians are likely to have an edge. Hence, if these claims withstand a test, it may simply be a function of the bias built into the formulation of the problem situation in the first place. Of course, one of the best ways to refute the principles of a particular school is to set them out in their own terms and then show that they are mistaken. If you can refute the views of a particular school using the vocabulary of that school, then you certainly have presented a strong refutation.

The people at Virginia Tech took another tack – setting out and classifying the principles to be tested as neutrally as possible. The hope was that any significant biases built into their terminology and evaluative procedures would be discovered in the course of the study. Of all the principles that they proposed to test, I will discuss only one – the role of novel predictions in science. Controversies in the English-speaking community over the role of novel predictions in science go back from Popper (1962), Hempel (1966), and Lakatos (1970), to at least Mill (1843) and Whewell (1849). The problem is not just that these philosophers meant different things by "novel predictions" but that they were not all that good about telling their readers what they actually meant by this phrase. One important distinction is between the phenomenon actually being observed (i.e., "known") and whether it follows from a particular theory. Four possibilities present themselves:

(i) Some phenomena are well known (they have been observed) and follow from one or more accepted theories. They are explained.
(ii) Some phenomena are not known (they have never been observed) but follow from a particular theory. These are epistemologically novel predictions.
(iii) Some phenomena are known to occur but have yet to be explained by current theories. They are curiosities.
(iv) Finally, some phenomena have not been experienced by anyone, and no theory implies that they should exist. They are unknown.

The situations relevant to our discussion are (ii) and (iii). In (ii) a phenomenon has yet to be noticed by the relevant scientists. Hence, it cannot contribute to the construction of a theory. However, once the theory has been sufficiently well constructed, it might well imply that such a phenomenon should occur. These scientists go check and *voilà*, they confirm this epistemologically novel prediction. In (iii) a phenomenon has been observed, but it has yet to be derived from any theory. Perhaps no theory currently exists from which it can be derived, or it could be the case that a theory exists, but no one has yet produced the derivation. It is derivable but not yet derived. After the fact, a scientist shows how this phenomenon can be derived from this new theory. The phenomenon is not novel, but the derivation is. This situation is termed "use novelty."

Why should the prediction of epistemologically novel phenomena be all that special? The usual answer turns on the avoidance of ad hoc hypotheses. If scientists are already aware of a particular phenomenon, they can build it into their theory from the start. Hence, no one is much impressed when they extract from a theory the very phenomenon that they build into it in the first place. However, if the scientists constructing a theory are not aware of this particular phenomenon, they cannot incorporate it into their theory in advance. Hence, the derivation of this unknown phenomenon and its later observation can be very persuasive.

This line of reasoning can be found in Hempel's discussion of novel predictions. According to Hempel (1966, p. 37), "it is highly desirable for a scientific hypothesis to be confirmed by 'new' evidence – by facts that were not known or not taken into account when the hypothesis was formed." But immediately thereafter, Hempel (1966, p. 38) brings himself up short:

> from a logical point of view, the strength of the support that a hypothesis receives from a given body of data should depend only on what the hypothesis asserts and what the data are: the question of whether the hypothesis or the data were presented first, being a purely historical matter, should not count as affecting the confirmation of the hypothesis.

In the first quotation, Hempel distinguishes between (ii) epistemological novelty (facts that were not known) and (iii) use novelty (facts that were not taken into account). According to the "positivist" school that Hempel represents, scientists might take the prediction of phenomena that have yet to be observed as providing greater support to a hypothesis than the explanation of phenomena that have already been observed, but they are mistaken to do so. Similar observations follow for use novelty. Either an observation is derivable from a set of premises or it is not, regardless of whether anyone has ever actually derived it. Science must be extracted from the contingencies of history and rationally reconstructed entirely in terms of inference. An observation statement follows from a set of premises or it does not, period. Attitudes such as this one are precisely what advocates of the historical school have attempted to combat.

However, when we turn to the work of Lakatos (1970) as a representative of the historical school, his views on this score are not as different from those of Hempel as one might expect.[2] Lakatos's methodology of scientific research programs is "historical" in the sense that the unit of evaluation is a temporal sequence of theories. Such a sequence of theories is theoretically progressive if each new theory in the sequence predicts "some novel, hitherto unexpected fact" (Lakatos 1970, p. 118), but as was his habit, Lakatos immediately transforms this apparently straightforward claim in counterintuitive ways. According to Lakatos (1970, p. 156), when a new research program explains "old facts" in novel ways, they count as "new facts"! In addition, all such appraisals are a matter of hindsight. It may take a long time before a research program can be "seen to produce 'genuinely novel' facts" (Lakatos 1970, p. 156). Unlike Hempel, Lakatos's scientific research programs have a temporal dimension but one that is crucially retrospective. Even determining which statements are part of a particular scientific research program is a function of hindsight. Lakatos's method of scientific research programs may be historical, but it is Whig history.

Before proceeding further, a word must be said about the "rational reconstructions" of which "positivist" philosophers of science are so fond, especially when we have emphasized all the reconstruction that must be carried out if hypotheses are to be tested empirically. The purpose of rational reconstructions is to make logical connections more apparent. Real science tends to be very messy, so messy that its logical structure is obscured. Rational reconstructions lay bare the logical structure by eliminating all historical, psychological, and sociological contingencies. The sort of reconstruction that I discussed previously involves formulating rough-and-ready operational criteria to aid in the testing of more general and abstract hypotheses. In this respect, these reconstructions are the exact opposite of rational reconstructions. They reintroduce the very contingencies that rational reconstructions are designed to exclude.

All of the preceding concerns what the principle of novel prediction actually asserts and whether it *should* play an important role in science. It is quite another question whether scientists actually *do* make recourse to this principle. The three authors in Donovan, Laudan, and Laudan (1988, pp. 18–20) who discuss this principle at any length all conclude that it did not play a very significant role in the scientific episodes that they studied (Hofmann 1988; Frankel 1988; and Nunan 1988).[3] Hence, the views of "positivists" on this score do not conflict with actual scientific practice. However, such coincidences do not function as confirmation for "positivists" because they do not think that philosophical views such as theirs can be tested empirically in the first place.

Initially, the function of epistemically novel predictions was to guard against the introduction of ad hoc hypotheses. If scientists give no special weight to novel predictions, then perhaps they are not as worried about ad hoc theorizing as they and students of science claim that they should be. Was the introduction of so many epicycles in Ptolemaic astronomy part of the legitimate articulation of this scientific theory or an instance of blatant ad hocery? How about the introduction of epistatic genes by Mendelian geneticists when transmission patterns did not accord with the basic principles of Mendelian genetics? Was the postulation of epistatic genes ad hoc? More often than not, claims about which hypotheses are ad hoc, and which are not, are largely a function of public relations. "My hypotheses are firmly grounded in careful observation, while yours are invented merely for the purposes at hand."

More fundamentally, claims about ad hoc hypotheses need not turn on the content of these claims but on a commitment to do something with them. A scientist can introduce what looks very much like an ad hoc hypothesis. If that is all that the scientist does with it, then it may well turn out to be ad hoc. However, if this scientist continues to work on the processes and entities postulated, developing and modifying them as he or she continues to work on the overall theory, then they may become a legitimate part of the theory. They may well have been introduced to neutralize a particular problem but were expanded to become an integral part of the theory, implying much more than the phenomenon that they were introduced to handle.

CONCLUSION

The main lessons to be learned from the preceding summaries of three different attempts to test hypotheses about science are that formulating hypotheses about science in ways that can be tested empirically is very difficult and that the necessary clarifications cannot be anticipated prior to our attempts at testing. One reason that novel predictions cannot play the clear-cut epistemological role that some have attributed to them is that the course of science does not run smoothly. On the basis of minimal data (some of which are likely to turn out to be mistaken or misleading) and very hazy theories, tests are run. Before these tests are completed, the scientists running them have already reformulated some of their hypotheses and made adjustments in their investigative techniques. Frequently, early test runs are not worth completing. Instead, new sequences of tests are devised and carried to partial completion, each more sophisticated and determinate than the ones before. Order is introduced only long after the fact, when it comes time to publish the results of one's research.

If scientists already knew everything that they needed to know before they started their empirical investigations, a single experiment might prove conclusive. Of course, scientists do not know in advance everything that they need to know. As a result, sequences of tests need to be run in order to make proper use of the ad hoc hypotheses that scientists formulate, and in the process some of these ad hoc hypotheses become transmuted into legitimate scientific hypotheses. Lakatos is right about this much. We can decide which hypotheses are ad hoc, and which are not, only in retrospect. In a genuinely historical view of science, many decisions can be made only in retrospect.

Parallel lessons hold for the study of science itself. I did my study on age in my spare time, with the help of two graduate students toward the end. (Incidentally, I did the data collecting because these graduate students did not know enough about the period under investigation to do it properly.) Sulloway was fortunate to be awarded a five-year grant from the MacArthur Foundation. In a paper that Rachel Laudan delivered at the History of Science Society meetings in 1990, she remarked on the inability of the people at Virginia Tech to get adequate funding for their project. The methods that Sulloway and I used did not require large research teams and massive financial expenditures. We could carry out our investigations in relative isolation for very little money, but if science teaches us anything about science, it implies that the investigation of certain problems requires large numbers of people working in close cooperation over many years. Haphazard, local, short-term investigations are not good enough. The people at Virginia Tech learned a lot about their research program as it progressed. By the time that the first cycle was complete, they knew enough to make the second cycle even more determinate. Unfortunately, no second cycle was possible. The funding ceased, and the members of this research team went their separate ways.

The conceptual obstacles confronting science studies are formidable. The disciplinary obstacles are even more formidable. No only are those of us who study science subdivided into numerous factions, each fighting to maintain its own turf, but also we do not have a tradition of the sort of funding necessary to make headway on the kinds of problems that we are addressing. It also does not help that science studies is seen by many scientists as a threat, a perception that is not totally without foundation. Many of those who study science are more interested in debunking it than in understanding it. The result has been "Science Wars."[4] Relativist students of science present numerous case studies to show how unimportant anything that might be termed evidence is in changing scientists' minds, but something is desperately wrong with presenting evidence to show how irrelevant evidence actually is. Now that students of science have retraced all the familiar ground that generations of philosophers have trod before them, perhaps we can abandon our intense fascination with philosophical puzzles and return to studying science.

NOTES

Reprinted with permission from *Perspectives on Science* 6 (1998), no. 3.
© 1999 by The Massachusetts Institute of Technology.
Special thanks are owed to Frank Sulloway and two referees who offered extensive and constructive suggestions for improving this paper.

1. The following authors discuss or at least mention Planck's principle: Lavoisier ([1777] 1862, 2:505); Whewell (1837, 2:157); Whewell (1851, p. 139); Comte (1853, 2:152); Darwin (1859, pp. 481–2); Lyell (1881, 2:253); Darwin (1899, 2:85, 218); Huxley (1901); Loewenberg (1932, p. 687); Planck (1936, p. 97); Planck (1949, pp. 33–4); Barber (1961, p. 596); Kuhn (1961, pp. 161–90); Kuhn (1962, pp. 89–90, 150–1); Oakley (1964); Hagstrom (1965, pp. 283, 291); Samuelson (1966, 2:1517–18); Greenberg (1967, p. 45); Scheffler (1967, pp. 18–19); Feyerabend (1970, p. 203); Dolby (1971, p. 19); Cole and Cole (1973, p. 82); Holton (1973, p. 394); Merton and Zuckermann (1973, pp. 497–559); Montgomery (1974, p. 115); Paul (1974, p. 412); Wisdom (1974, 2:829); Bondi (1975, p. 7); Cantor (1975, p. 196); Cole (1975, p. 181); Gunther (1975, p. 458); Brush (1976, p. 640); Chubin (1976, p. 464); Garber (1976, p. 96); Knight (1976, p. 14); Stegmuller (1976, p. 148); Rosenkrantz (1977, p. 251); Backmore (1978, pp. 347–9); Blua (1978, p. 197); Hull, Tessner, and Diamond (1978, pp. 717–23); McCann (1978, pp. 21, 34, 61–2, 79–80, 90–1, 101, 119); Nitecki, Lemke, Pullman, and Johnson (1978, pp. 661–4); Barash (1979, p. 240); Garvey (1979, p. 15); Hufbauer (1979, p. 744); Mahoney (1979, p. 372); Coats (1980, pp. 190–2); Diamond (1980, pp. 838–41); Newton-Smith (1981, p. 235); Broad and Wade (1982, p. 135); Cock (1983, p. 40); Grayson (1983, p. 208); Messeri (1988, pp. 91–112); Perrin (1988, pp. 105–24); Darwin (1991, p. 279); Stephan and Levin (1992); Margolis (1996, p. 136).

2. The classificatory criteria that I am using in placing particular philosophers in particular schools are extremely crude and not very faithful to history. In many important respects, Hempel is anything but a positivist, and Lakatos retains many positivistic elements.

3. One feature of empirical testing is that results may conflict with one's own preferences. For example, I think that novel predictions should lend greater support to a hypothesis than the derivation of the commonplace. I also thought that practicing scientists have more sense than their positivist commentators and treat novel predictions as special. Unfortunately, on the basis of these three studies, I am forced to admit that I was wrong.

4. Several publications have really heated up the war between the relativist, deconstructionist critics of science and their realist, "positivist" opponents: Gross and Levitt (1994); Gross, Levitt, and Lewis (1996); and Sokal (1996a, 1996b). Gross, Levitt, and Lewis assay what they take to be left-wing, deconstructionist, postmodernist critics of science, while Sokal wrote a parody of the sort of prose that he thought characterizes this same literature and got it published, not as a parody but as a genuine contribution. The reactions have been numerous and heated: see Richards and Ashmore (1996); Fish (1996); Fuller (1997); Dickson (1997a, 1997b); Gottfried and Wilson (1997); Macilwain (1997a, 1997b); Gross (1997); Edge (1997); Forman (1997a, 1997b);

Robinson (1997); Levitt (1997); Trefil (1997); Herschbach (1997); Sandler (1997); Ziegler (1997); and Gibbons (1997). One characteristic of the various defenses of the new science studies is that these defenders are rapidly taking back everything that was new about the "new" science studies. No matter how crystal clear their prose may have been, their critics have totally misconstrued them as saying something novel and interesting.

References

Akeroyd, F. M. 1991. A Practical Example of Grue. *British Journal for the Philosophy of Science* 42:535–7.

Alexander, R. D. 1979. *Darwinism and Human Affairs*. Seattle: University of Washington Press.

Andersson, D. I., Slechta, E. S., and Roth, J. R. 1998. Evidence That Gene Amplification Underlies Adaptive Mutability of the Bacterial Lac Operon. *Science* 282:1133–5.

Antonovics, J., Ellstrand, N. C., and Brandon, R. N. 1988. Genetic Variation and Environmental Variation: Expectations and Experiments. In *Plant Evolutionary Biology*, ed. L. D. Gottlieb and S. K. Jain. London: Chapman and Hall, pp. 275–303.

Arnold, A. J., and Fristrup, K. 1981. The Theory of Evolution by Natural Selection: A Hierarchical Expansion. *Paleontology* 8:113–29.

Atran, S. 1990. *Cognitive Foundations of Natural History: Towards an Anthropology of Science*. Cambridge University Press.

Babchuck, N., and Bates, A. P. 1962. Professor or Producer: The Two Faces of Academic Man. *Social Forces* 40:341–8.

Bacon, F. [1620] 1960. *The New Organon and Related Writings*, ed. F. H. Anderson. Indianapolis: Bobbs-Merrill.

Barash, D. 1979. *The Whisperings Within*. New York: Penguin.

Barber, B. 1961. Resistance by Scientists to Scientific Discovery. *Science* 134:596–602.

Barnes, B. 1977. *Interests and the Growth of Knowledge*. London: Routledge and Kegan Paul.

Benson, S. 1997. Adaptive Mutation: A General or Special Case? *Bioessays* 19:9–11.

Berlin, B., and Kay, P. 1969. *Basic Color Terms*. Berkeley: University of California Press.

Blackmore, J. 1978. Is Planck's "Principle" True? *British Journal for the Philosophy of Science* 29:347–9.

Blau, J. R. 1978. Sociometric Structure of a Scientific Discipline. *Research in Sociology of Knowledge, Sciences and Art* 1:191–206.

Bliss, M. 1982. *The Discovery of Insulin*. Chicago: University of Chicago Press.

References

Bloor, D. 1976. *Knowledge and Social Imagery*. London: Routledge and Kegan Paul.

1997. Obituary of Thomas Kuhn. *Social Studies of Science* 27:498–502.

Blough, D. S. 1963. Interresponse Function of Continuous Variables: A New Method and Some Data. *Journal of the Experimental Analysis of Behavior* 6:237–46.

Bondi, H. 1975. What Is Progress in Science? In *Problems of Scientific Revolutions*, ed. R. Harré. Oxford: Clarendon Press, pp. 1–10.

Bonner, J. T. 1974. *On Development*. Cambridge, MA: Harvard University Press.

Bornstein, H. H. 1973. Color Vision and Color Naming: A Psychophysiological Hypothesis of Cultural Difference. *Psychological Bulletin* 80:257–87.

Boyd, R., and Richerson, P. J. 1985. *Culture and the Evolutionary Process*. Chicago: University of Chicago Press.

1989. The Evolution of Indirect Reciprocity. *Social Networks* 11:213–36.

Bradie, M. 1986. Assessing Evolutionary Epistemology. *Biology & Philosophy* 1:401–60.

1991. The Evolution of Scientific Lineages. In *PSA 1990*, vol. 2, ed. A. Fine, M. Forbes, and L. Wessels. East Lansing, MI: Philosophy of Science Association, pp. 245–54.

Brandon, R. N. 1982. The Levels of Selection. In *PSA 1980*, vol. 2, ed. P. D. Asquith and T. Nickles. East Lansing, MI: Philosophy of Science Association, pp. 315–23.

1990. *Adaptation and Environment*. Princeton: Princeton University Press.

Brandon, R. N., Antonovics, J., Burian, R., Carson, S., Cooper, G., Davies, P. S., Horvath, C., Mishler, B. D., Richardson, R. C., Smith, K., and Thrall, P. 1994. Sober on Brandon on Screening-off and the Levels of Selection. *Philosophy of Science* 61:475–86.

Brandon, R. N., and Burian, R. N., eds. 1984. *Genes, Organisms, Populations*. Cambridge, MA: MIT Press.

Broad, W., and Wade, N. 1982. *Betrayers of the Truth*. New York: Simon and Schuster.

Brookfield, J. F. Y. 1995. Evolving Darwinism. *Nature* 376:551–2.

Brooks, D. R., and Wiley, E. O. 1986. *Evolution and Entropy*. Chicago: University of Chicago Press.

Brown, J. R. 1991. *The Laboratory of the Mind: Thought Experiments in the Natural Sciences*. London: Routledge and Kegan Paul.

1993. Why Empiricism Won't Work. In *PSA 1992*, vol. 2, ed. D. Hull, M. Forbes, and K. Okruhlik. East Lansing, MI: Philosophy of Science Association, pp. 271–9.

Brush, S. G. 1976. *The Kind of Motion We Call Heat*. New York: American Elsevier.

1989. Prediction and Theory Evaluation: The Case of Light Bending. *Science* 246:1124–9.

1993. Prediction and Theory Evaluation: Cosmic Microwaves and the Revival of the Big Bang. *Perspectives on Science* 1:515–602.

Burkhardt, F. 1974. England and Scotland: The Learned Societies. In *The Com-*

parative Reception of Darwinism, ed. T. F. Glick. Austin: University of Texas Press, pp. 32–74.

Campbell, D. T. 1974. Evolutionary Epistemology. In *The Philosophy of Karl R. Popper*, ed. P. A. Schilpp. La Salle, IL: Open Court, pp. 413–63.

——— 1979. A Tribal Model of the Social System Vehicle Carrying Scientific Knowledge. *Knowledge: Creation, Diffusion, Utilization* 1:181–201.

Campbell, D. T., Heyes, C. M., and Callebaut, W. B. 1987. Evolutionary Epistemology: A Bibliography. In *Evolutionary Epistemology: A Multiparadigm*, ed. W. Callebaut and R. Pinxten. Dordrecht, Holland: Reidel, pp. 401–31.

Cantor, G. N. 1975. The Edinburgh Phrenology Debate: 1803–1828. *Annals of Science* 32:195–218.

Caplan, A. 1980. Have Species Become Déclassé? In *PSA 1980*, vol. 1, ed. P. D. Asquith and R. N. Giere. East Lansing, MI: Philosophy of Science Association, pp. 71–82.

——— 1981. Back to Class: A Note on the Ontology of Species. *Philosophy of Science* 48:130–40.

Catania, A. C. 1973. The Concept of the Operant in the Analysis of Behavior. *Behaviorism* 1973:103–16.

Cavalli-Sforza, L., and Feldman, M. 1981. *Cultural Transmission and Evolution: A Quantitative Approach*. Princeton: Princeton University Press.

Chubin, D. E. 1976. The Conceptualization of Scientific Specialties. *Sociological Quarterly* 17:448–76.

Churchland, P. 1982. Is the Thinker a Natural Kind? *Dialogue* 21:223–38.

Close, F. 1990. *Too Hot to Handle: The Race for Cold Fusion*. Princeton: Princeton University Press.

Coats, A. W. 1980. The Historical Context of the "New" Economic History. *Journal of European Economic History* 9:185–207.

Cock, A. G. 1983. William Bateson's Rejection and Eventual Acceptance of Chromosome Theory. *Annals of Science* 40:19–60.

Cole, J., and Cole, S. 1972. The Ortega Hypothesis. *Science* 178:368–75.

——— 1973. *Social Stratification in Science*. Chicago: University of Chicago Press.

Cole, S. 1970. Professional Standing and the Reception of Scientific Discoveries. *American Journal of Sociology* 76:286–306.

——— 1975. The Growth of Scientific Knowledge. In *The Idea of Social Structure*, ed. L. A. Coser. New York: Harcourt, Brace, Jovanovich, pp. 175–220.

——— 1991. *Social Influences on Science*. Cambridge, MA: Harvard University Press.

——— 1992. *Making Science: Between Nature and Society*. Cambridge, MA: Harvard University Press.

Cole, S., Cole, J. R., and Simon, G. S. 1981. Chance and Consensus in Peer Review. *Science* 213:881–6.

Cole, S., Rubin, L., and Cole, J. R. 1978. *Peer Review in the National Science Foundation: Phase I*. Washington, D.C.: National Academy of Sciences.

Collier, J. 1990. Could I Conceive of Being a Brain in a Vat? *Australasian Journal of Philosophy* 68:413–19.

References

Collins, H. M. 1975. The Seven Sexes: A Study in the Sociology of a Phenomenon, or the Replication of Experiments in Physics. *Sociology* 9:205–44.

1981a. What Is TRASP?: The Radical Programme as a Methodological Imperative. *Philosophy of the Social Sciences* 11:215–24.

1981b. Stages in the Empirical Program of Relativism. *Social Studies of Science* 11:3–10.

1985. *Changing Order: Replication and Induction in Scientific Practice*. London: Sage.

Comte, A. 1853. *The Positive Philosophy*. London: John Chapman.

Connor, S. 1987. AIDS: Science Stands on Trial. *New Scientist* 113:49–58.

Cook, R. E. 1980. Asexual Reproduction: A Further Consideration. *American Naturalist* 113:769–72.

Cracraft, J. 1989. Species as Entities of Biological Theory. In *What the Philosophy of Biology Is: Essays Dedicated to David Hull*, ed. M. Ruse. Dordrecht, Holland: Kluwer, pp. 31–52.

Crow, F. F. 1999. Unmasking a Cheating Gene. *Science* 283:1651–2.

Culliton, B. 1983. Coping with Fraud: The Darsee Case. *Science* 220:31–5.

1986. Harvard Researchers Retract Data in Immunology Paper. *Science* 234:1069.

1987. Integrity of Research Papers Questioned. *Science* 235:422–3.

Cziko, G. 1995. *Without Miracles: Universal Selection Theory and the Second Darwinian Revolution*. Cambridge, MA: MIT Press.

Dancy, J. 1985. The Role of Imaginary Cases in Ethics. *Pacific Philosophical Quarterly* 66:141–53.

Darden, L., and Cain, A. J. 1989. Selection Type Theories. *Philosophy of Science* 56:106–29.

Darwin, C. [1859] 1964. *On the Origin of Species*. Cambridge, MA: Harvard University Press.

1881. *The Formation of Vegetable Mould, through the Action of Worms, with Observations on their Habits*. London: Murray.

1899. *The Life and Letters of Charles Darwin*. New York: Appleton.

1903. *More Letters of Charles Darwin*. New York: Appleton.

1991. *The Correspondence of Charles Darwin*, ed. F. Burckhardt and S. Smith. Cambridge, MA: Cambridge University Press.

Davidson, E. H., Peterson, K. J., and Cameron, R. A. 1995. Origin of Bilateral Body Plans: Evolution of Developmental Regulatory Mechanisms. *Science* 270:1319–25.

Dawkins, R. 1976. *The Selfish Gene*. Oxford: Oxford University Press.

1982a. Replicators and Vehicles. In *Current Problems in Sociobiology*, ed. King's College Sociobiology Group. Cambridge University Press, pp. 45–64.

1982b. *The Extended Phenotype*. San Francisco: Freeman.

1983. Universal Darwinism. In *Evolution from Molecules to Men*, ed. D. S. Bendall. Cambridge University Press, pp. 403–25.

1994. Burying the Vehicle. *Behavioral and Brain Sciences* 17:616–17.

1996. Reply to Phillip Johnson. *Biology & Philosophy* 1:539–40.

Dennett, D. 1995. *Darwin's Dangerous Idea*. New York: Simon and Schuster.

References

Desmond, A. 1989. *The Politics of Evolution: Morphology, Medicine, and Reform in Radical London*. Chicago: University of Chicago Press.

Diamond, A. M. 1980. Age and Acceptance of Cliometrics. *Journal of Economic History* 40:838–41.

1997. The Economics of Science. *Knowledge and Policy* 9:3–142.

Dickson, D. 1997a. The "Sokal Affair" Takes a Transatlantic Turn. *Nature* 385:381.

1997b. Champions or Challengers of the Cause of Science? *Nature* 387:333–4.

Dobzhansky, T. [1937] 1951. *Genetics and the Origin of Species*, 3d ed. New York: Columbia University Press.

Dolby, R. G. A. 1971. Sociology of Knowledge in Natural Science. *Science Studies* 1:3–21.

Donahoe, J. W., and Palmer, D. C. 1994. *Learning and Complex Behavior*. Boston: Allyn and Bacon.

Donoghue, M. J. 1990. Sociology, Selection, and Success: A Critique of David Hull's Analysis of Science and Systematics. *Biology & Philosophy* 5:459–72.

Donovan, A., Laudan, L., and Laudan, R., eds. 1988. *Scrutinizing Science*. Dordrecht, Holland: Kluwer.

Dretske, F. 1981. *Knowledge and the Flow of Information*. Cambridge, MA: MIT Press.

Dupré, J. 1990. Scientific Pluralism and the Plurality of Sciences: Comments on David L. Hull's *Science as a Process*. *Philosophical Studies* 69:61–76.

1993. *The Disunity of Things: Metaphysical Foundations of the Disunity of Science*. Cambridge, MA: Harvard University Press.

Edelman, G. M. 1987. *Neural Darwinism: The Theory of Neuronal Group Selection*. New York: Basic Books.

Edge, D. 1997. Evolution Teaching. *Science* 274:904.

Eldredge. N. 1985. *Unfinished Synthesis*. Oxford: Oxford University Press.

Eldredge, N., and Gould, S. J. 1972. Punctuated Equilibria: An Alternative to Phyletic Gradualism. In *Models in Paleontology*, ed. T. J. M. Schopf. San Francisco: Freeman and Cooper, pp. 82–115.

Eldredge, N., and Grene, M. 1992. *Interactions: The Biological Context of Social Systems*. New York: Columbia University Press.

Eldredge, N., and Salthe, S. N. 1984. Hierarchy and Evolution. *Oxford Surveys in Evolutionary Biology* 1:182–206.

Enç, B. 1995. Units of Behavior. *Philosophy of Science* 62:523–42.

Endler, J. A. 1986. *Natural Selection in the Wild*. Princeton: Princeton University Press.

Fernholm, B., Brener, K., and Jörnvall, H., eds. 1989. *The Hierarchy of Life: Molecules and Morphology in Phylogenetic Analysis*. Amsterdam: Elsevier.

Ferster, C. B., and Skinner, B. F. 1957. *Schedules of Reinforcement*. New York: Prentice Hall.

Feyerabend, P. 1970. Consolations for the Specialist. In *Criticism and the Growth of Knowledge*, ed. I. Lakatos and A. Musgrave. Cambridge University Press, pp. 197–230.

Feynman, R. P., Leighton, R. B., and Sands, M. 1964. *The Feynman Lectures on Physics*, vol. 2. Reading, MA: Addison-Wesley.

Fish, S. 1996. Professor Sokal's Bad Joke. *New York Times*. May 21, op-ed.

References

Fodor, J. 1964. On Knowing What We Would Say. *Philosophical Review* 73:198–212.

Forge, J. 1991. Thought Experiments in the Philosophy of Physical Science. In *Thought Experiments in Science and Philosophy*, ed. T. Horowitz and G. J. Massey. Savage, MD: Rowman and Littlefield, pp. 209–22.

Forman, P. 1997a. Assailing the Seasons. *Science* 276:750–2.

1997b. Deconstructing Science. *Science* 276:1955.

Fox, M. F. 1983. Publication Productivity among Scientists: A Critical Review. *Social Studies of Science* 13:285–305.

Frank, R. H. 1985. *Choosing the Right Pond: Human Behavior and the Quest for Status*. New York: Oxford University Press.

Frank, R. H., and Cook, P. 1995. *The Winner-Take-All Society*. New York: Free Press.

Frankel, H. 1988. Plate Tectonics and Inter-Theory Relations. In *Scrutinizing Science: Empirical Studies of Scientific Change*, ed. A. Donovan, L. Laudan, and R. Laudan. Dordrecht, Holland: Kluwer, pp. 269–87.

Fuller, S. 1994. Toward a Philosophy of Science Accounting: A Critical Rendering of Instrumental Rationality. *Science in Context* 7:591–621.

1997. Out of Context. *Nature* 385:109.

Gale, R. M. 1991. On Some Pernicious Thought-Experiments. In *Thought Experiments in Science and Philosophy*, ed. T. Horowitz and G. J. Massey. Savage, MD: Rowman and Littlefield, pp. 297–303.

Gally, J. A., and Edelman, G. M. 1972. The Genetic Control of Immunoglobulin Synthesis. *Annual Review of Genetics* 6:1–46.

Garber, E. 1976. Some Reactions to Planck's Law, 1900–1914. *Studies in History and Philosophy of Science* 7:89–126.

Garvey, W. D. 1979. *Communication: The Essence of Science*. New York: Pergamon.

Geertz, C. 1973. *The Interpretation of Cultures*. New York: Basic Books.

Ghiselin, M. T. 1966. On Psychologism in the Logic of Taxonomic Principles. *Systematic Zoology* 15:207–15.

1969. *The Triumph of the Darwinian Method*. Berkeley: University of California Press.

1974. A Radical Solution to the Species Problem. *Systematic Zoology* 23:536–44.

1981. Categories, Life, and Thinking. *Behavioral and Brain Sciences* 4:269–313.

1985. Species Concepts, Individuality, and Objectivity. *Biology & Philosophy* 2:127–45.

Gibbons, A. 1997. Cultural Divide at Stanford. *Science* 276:1783.

Giere, R. 1984. *Understanding Science*, 2d ed. New York: Holt, Rinehart, and Winston.

1988. *Explaining Science: A Cognitive Approach*. Chicago: University of Chicago Press.

Gilbert, T. 1958. Fundamental Dimensional Properties of the Operant. *Psychological Review* 65:272–82.

Glenn, S. S. 1991. Contingencies and Metacontingencies: Relations among Behavioral, Cultural, and Biological Evolution. In *Behavioral Analysis of*

Societies and Cultural Practices, ed. P. A. Lamal. Bristol, PA: Hemisphere, pp. 39–73.

Glenn, S. S., Ellis, J., and Greenspoon, J. 1992. On the Revolutionary Nature of the Operant as a Unit of Behavioral Selection. *American Psychologist* 47:1329–36.

Glenn, S. S., and Field, D. P. 1994. Functions of the Environment in Behavioral Evolution. *Behavior Analyst* 17:241–59.

Glenn, S. S., and Madden, G. J. 1995. Units of Interaction, Evolution, and Replication: Organic and Behavioral Parallels. *Behavior Analyst* 18:237–51.

Glymour, B. 1999. Population Level Causation and a Unified Theory of Natural Selection. *Biology & Philosophy* 14:521–36.

Goldman, A. 1988. *Empirical Knowledge*. Berkeley: University of California Press.

Goldman, A., and Shaked, M., eds. 1992. *Liaisons: Philosophy Meets the Cognitive and Social Sciences*. Cambridge, MA: MIT Press.

Gooding, D. 1990. *Experiment and the Making of Meaning*. Dordrecht, Holland: Kluwer.

1993. What Is "Experimental" about Thought Experiments? In *PSA 1992*, vol. 2, ed. D. Hull, M. Forbes, and K. Okruhlik. East Lansing, MI: Philosophy of Science Association, pp. 280–90.

Goodman, N. [1954] 1983. *Fact, Fiction, and Forecast*, 2d. ed. Cambridge, MA: Harvard University Press.

Gorham, G. 1991. Planck's Principle and Jean's Conversion. *Studies in History and Philosophy of Science* 22:471–91.

Gottfried, K., and Wilson, K. G. 1997. Science as a Cultural Construct. *Nature* 386:545–7.

Gould, S. J. 1977. Caring Groups and Selfish Genes. *Natural History* 86:20–4.

1997. Evolution: The Pleasures of Pluralism. *New York Review of Books*, June 26, pp. 47–52.

Gould, S. J., and Lewontin, R. C. 1979. The Spandrels of San Marco and the Panglossian Paradigm: A Critique of the Adaptational Programme. *Proceedings of the Royal Society of London* 205:581–98.

Gould, S. J., and Lloyd, E. A. Forthcoming. The Allometry of Individuality and Adaptation Across Levels of Selection: How Shall We Name and Generalize the Unit of Darwinism?

Gray, R. D. 1992. Death of the Gene: Developmental Systems Fight Back. In *Trees of Life: Essays in Philosophy of Biology*, ed. P. E. Griffiths. Oxford: Oxford University Press, pp. 165–209.

Grayson, D. K. 1983. *The Establishment of Antiquity*. New York: Academic Press.

Greenberg, D. S. 1967. *The Politics of Pure Science*. New York: New American Library.

Grene, M. 1987. Hierarchies in Biology. *American Scientist* 75:504–10.

1989. Interaction and Evolution. In *What the Philosophy of Biology Is: Essays Dedicated to David Hull*, ed. M. Ruse. Dordrecht, Holland: Kluwer, pp. 67–73.

Griesemer, J. 1998. The Case for Epigenetic Inheritance in Evolution. *Journal of Evolutionary Biology* 11:193–200.

1999. Materials for the Study of Evolutionary Transition. *Biology & Philosophy* 14:127–42.

References

Griesemer, J. R., and Wimsatt, W. 1989. Picturing Weismannism. In *What the Philosophy of Biology Is: Essays Dedicated to David Hull*, ed. M. Ruse. Dordrecht, Holland: Kluwer, pp. 75–137.

Griffiths, P. E., and Gray, R. D. 1994. Developmental Systems and Evolutionary Explanation. *Journal of Philosophy* 91:277–304.

Grinnell, F. 1993. Industrial Sponsors and the Scientist. *Journal of NIH Research* 5:50.

Gross, P. R. 1997. Characterizing Scientific Knowledge. *Science* 275:143.

Gross, P. R., and Levitt, N. 1994. *Higher Superstition*. Baltimore: Johns Hopkins University Press.

Gross, P. R., Levitt, N., and Lewis, M. W. 1996. *The Flight from Science and Reason*. New York: New York Academy of Sciences.

Gunther, A. E. 1975. *A Century of Zoology*. New York: Science History.

Hacking, I. 1993. Do Thought Experiments Have a Life of Their Own? Comments on James Brown, Nancy Nersessian, and David Gooding. In *PSA 1992*, vol. 2, ed. D. Hull, M. Forbes, and K. Okruhlik. East Lansing, MI: Philosophy of Science Association, pp. 302–8.

Hagstrom, W. 1965. *The Scientific Community*. New York: Basic Books.

Hamilton, A., Madison, J., and Jay, J. [1788] 1818. *The Federalist Papers on the New Constitution*. Washington, D. C.: Jacob Gideon.

Hamilton, D. 1990. Publishing by – and for – the Numbers? *Science* 250: 1331–2.

  1991a. NIH Finds Fraud in "Cell" Paper. *Science* 251:1552–4.

  1991b. Research Papers: Who's Uncited Now? *Science* 251:25.

Hamilton, W. D. 1964a. The Genetical Evolution of Social Behavior, I. *Journal of Theoretical Biology* 7:1–16.

  1964b. The Genetical Evolution of Social Behavior, II. *Journal of Theoretical Biology* 7:17–32.

Hardin, G. 1977. *The Limits of Altruism: An Ecologist's View of Survival*. Bloomington: Indiana University Press.

Harms, W. F. 1998. The Use of Information Theory in Epistemology. *Philosophy of Science* 65:472–501.

Harper, J. L. 1977. *Population Biology of Plants*. London: Academic Press.

Harré, R. 1979. *Social Being*. Oxford: Blackwell.

Hempel, C. G. 1966. *Philosophy of Natural Science*. Englewood Cliffs, NJ: Prentice Hall.

Hennig, W. 1966. *Phylogenetic Systematics*. Champaign: University of Illinois Press.

  1969. *Die Stammesgeschichte der Insectin*. English edition, 1981. *Insect Phylogeny*. New York: Wiley.

Herschbach, D. 1997. Deconstructing Science. *Science* 276:1954.

Herschel, J. F. W. [1841] 1987. *A Preliminary Discourse on the Study of Natural Philosophy*. Chicago: University of Chicago Press.

Heschel, A. 1994. Reconstructing the Real Unit of Selection. *Behavioral and Brain Sciences* 174:624–5.

Heyes, C. 1988. Are Scientists Agents in Scientific Change? *Biology & Philosophy* 3:194–9.

References

Hinton, G. E., and Nowland, S. J. 1987. How Learning Can Guide Evolution. *Complex Systems* 1:495–502.

Hirshleifer, J. 1977. Economics from a Biological Viewpoint. *Journal of Law and Economics* 20:1–52.

Hodge, J. 1989. Darwin's Theory and Darwin's Argument. In *What the Philosophy of Biology Is: Essays Dedicated to David Hull*, ed. M. Ruse. Dordrecht, Holland: Kluwer, pp. 163–82.

Hofmann, J. R. 1988. Ampère's Electrodynamics and the Acceptability of Guiding Assumptions. In *Scrutinizing Science: Empirical Studies of Scientific Change*, ed. A. Donovan, A. L. Laudan, and R. Laudan. Dordrecht, Holland: Kluwer, pp. 201–17.

Holton, G. 1973. *Thematic Origins of Scientific Thought*. Cambridge, MA: Harvard University Press.

Horowitz, T. 1991. Newcomb's Problem as a Thought Experiment. In *Thought Experiments in Science and Philosophy*, ed. T. Horowitz and G. J. Massey. Savage, MD: Rowman and Littlefield, pp. 305–16.

Horowitz, T., and Massey, G. J. 1991. Introduction. In *Thought Experiments in Science and Philosophy*, ed. T. Horowitz and G. J. Massey. Savage, MD: Rowman and Littlefield, pp. 1–26.

Hufbauer, K. 1979. A Test of the Kuhnian Theory. *Science* 20:744–5.

Hull, D. L. 1972. Reduction in Genetics – Biology or Philosophy? *Philosophy of Science* 39:491–9.

  1973. Reduction in Genetics – Doing the Impossible. In *Proceedings of the Fourth International Congress of Logic, Methodology and Philosophy of Science*, ed. P. Suppes. Amsterdam: North-Holland, pp. 619–35.

  1976. Are Species Really Individuals? *Systematic Zoology* 25:174–91.

  1978a. A Matter of Individuality. *Philosophy of Science* 45:335–60.

  1978b. Altruism in Science: A Sociobiological Model of Cooperative Behavior among Scientists. *Animal Behaviour* 26:685–97.

  1980. Individuality and Selection. *Annual Review of Ecology and Systematics* 11:311–32.

  1981a. Kitts and Kitts and Caplan on Species. *Philosophy of Science* 48:141–52.

  1981b. Units of Evolution: A Metaphysical Essay. In *The Philosophy of Evolution*, ed. U. J. Jensen and R. Harré. Brighton: Harvester, pp. 23–44.

  1982. The Naked Meme. In *Learning, Development, and Culture*, ed. H. C. Plotkin. New York: Wiley.

  1983a. Thirty-One Years of "Systematic Zoology." *Systematic Zoology* 32:315–42.

  1983b. Exemplars and Scientific Change. In *PSA 1982*, vol. 2., ed. P. D. Asquith and T. Nickles. East Lansing, MI: Philosophy of Science Association, pp. 479–503.

  1984a. Cladistic Theory: Hypotheses that Blur and Grow. In *Cladistic Perspectives on the Reconstruction of Evolutionary History*, ed. T. Duncan and T. Stuessy. New York: Columbia University Press, pp. 5–23.

  1984b. Lamarck among the Anglos. Introduction to J. B. Lamarck, *Zoological Philosophy* (1809). Chicago: University of Chicago Press.

References

1985a. Darwinism as a Historical Entity: A Historiographic Proposal. In *The Darwinian Heritage*, ed. D. Kohl. Princeton: Princeton University Press.

1985b. Bias and Commitment in Science: Phenetics and Cladistics. *Annals of Science* 42:319–38.

1985c. Openness and Secrecy in Science: Their Origins and Limitations. *Science, Technology, and Human Values* 10:4–13.

1986. Conceptual Evolution and the Eye of the Octopus. In *Logic, Methodology and Philosophy of Science*, ed. R. B. Marcus, G. J. W. Dorn, and P. Weingartner. Amsterdam: North-Holland, pp. 643–65.

1987. Genealogical Actors in Ecological Plays. *Biology & Philosophy* 1:44–60.

1988a. *Science as a Process: An Evolutionary Account of the Social and Conceptual Development of Science*. Chicago: University of Chicago Press.

1988b. A Mechanism and Its Metaphysics: An Evolutionary Account of the Social and Conceptual Development of Science. *Biology & Philosophy* 3:123–55, and author's response, 3:241–63.

1988c. Interactors versus Vehicles. In *The Role of Behavior in Evolution*, ed. H. C. Plotkin. Cambridge, MA: MIT Press, pp. 19–50.

1989a. A Function for Actual Examples in Philosophy of Science. In *What the Philosophy of Biology Is: Essays Delicated to David Hull*, ed. M. Ruse. Dordrecht, Holland: Kluwer, pp. 313–24.

1989b. *The Metaphysics of Evolution*. New York: State University of New York Press.

1990. Particularism in Science. *Criticism* 32:343–59.

1991. Conceptual Evolution: A Response. In *PSA 1990*, vol. 2, ed. A. Fine, M. Forbes, and L. Wessels. East Lansing, MI: Philosophy of Science Association, pp. 255–64.

1994. Taking Vehicles Seriously. *Behavioral and Brain Sciences* 17:627–8.

Hull, D. L., Tessner, Peter, and Diamond, Arthur. 1978. Planck's Principle. *Science* 202:717–23.

Huxley, T. H. 1901. *Life and Letters of Thomas Henry Huxley*. New York: Appleton.

Irvine, A. 1991. On the Nature of Thought Experiments in Scientific Reasoning. In *Thought Experiments in Science and Philosophy*, ed. T. Horowitz and G. J. Massey. Savage, MD: Rowman and Littlefield, pp. 49–165.

Jablonka, E., and Lamb, M. 1995. *Epigenetic Inheritance and Evolution*. Oxford: Oxford University Press.

Jackson, J. B. D., Buss, L. W., and Cook, R. E., eds. 1986. *Population Biology and Evolution of Clonal Organisms*. New Haven: Yale University Press.

Janzen, D. W. 1977. What Are Dandelions and Aphids? *American Naturalist* 111:586–9.

Kawata, M. 1987. Units and Passages: A View for Evolutionary Biology. *Biology & Philosophy* 2:425–34.

Kevles, D. J. 1996. The Assault on David Baltimore. *New Yorker*, May 27, pp. 94–109.

Kim, K. 1994. *Explaining Scientific Consensus*. New York: Guilford.

Kimura, M. 1983. *The Neutral Theory of Molecular Evolution*. Cambridge University Press.

References

King, J. L. 1984. Selectively Neutral Alleles with Significant Phenotypic Effects: A Steady-State Model. *Evolutionary Theory* 7:73–9.

Kitcher, P. 1978. Theories, Theorists and Theoretical Change. *Philosophical Review* 87:519–47.

——— 1984. Against the Monism of the Moment: A Reply to Elliott Sober. *Philosophy of Science* 51:616–30.

——— 1989. Some Puzzles about Species. In *What the Philosophy of Biology Is: Essays Dedicated to David Hull*, ed. M. Ruse. Dordrecht, Holland: Kluwer, pp. 183–208.

——— 1992. Authority, Deference, and the Role of Individual Reason. In *The Social Dimension of Science*, ed. E. McMullin. Notre Dame, IN: University of Notre Dame Press, pp. 244–71.

——— 1993. *The Advancement of Science*. New York: Oxford University Press.

Kitts, D. B., and Kitts, D. J. 1979. Biological Species as Natural Kinds. *Philosophy of Science* 46:613–22.

Knight, D. 1976. *The Nature of Science*. London: A. Deutsch.

Koshland, D. E. 1987. Fraud in Science. *Science* 235:197.

Kripke, S. 1972. Naming and Necessity. In *Semantics and Natural Languages*, ed. D. Davidson and G. Harman. Dordrecht, Holland: Reidel, pp. 235–355.

Kuhn, T. S. 1961. The Function of Measurement in Modern Physical Theories. *Isis* 52:161–90. Reprinted (1977) in *The Essential Tension*. Chicago: University of Chicago Press, pp. 178–224.

——— [1962] 1970. *The Structure of Scientific Revolutions*, 2d ed. Chicago: University of Chicago Press.

——— 1964. A Function for Thought Experiments. *L'aventure de la Science, Mélanges Alexandre Koyré*, Vol. 2 Paris: Hermann, pp. 307–34. Reprinted (1977) in *The Essential Tension*. Chicago: University of Chicago Press, pp. 240–65.

Küppers, B-O. 1990. *Information and the Origin of Life*. Cambridge, MA: MIT Press.

Ladd, E. C., and Lipset, S. M. 1977. Survey of 4,400 Faculty Members at 161 Colleges and Universities. *Chronicle of Higher Education*, November 21, p. 12; November 28, p. 2.

Lakatos, I. 1970. Falsification and the Methodology of Scientific Research Programmes. In *Criticism and the Growth of Knowledge*, ed. I. Lakatos and A. Worrall. Cambridge University Press, pp. 91–195.

——— 1971. History of Science and Its Rational Reconstruction. In *Boston Studies in the Philosophy of Science*, vol. 8, ed. R. C. Buck and R. S. Cohen. Dordrecht, Holland: Reidel, pp. 91–136.

Laland, K. N., Odling-Smee, J., and Feldman, M. W. 2000. Niche Construction, Biological Evolution and Cultural Change. *Behavioral and Brain Sciences* 23.

Langman, R. 1989. *The Immune System*. New York: Academic Press.

Langman, R., and Cohn, M. 1996. A Short History of Time and Space in Immune Discrimination. *Scandinavian Journal of Immunology* 4:544–8.

Laudan, L. 1989. The Rational Weight of the Scientific Past: Forging Fundamental Change in a Conservative Discipline. In *What the Philosophy of Biology Is: Essays Dedicated to David Hull*, ed. M. Ruse. Dordrecht, Holland: Kluwer, pp. 209–20.

References

Lavoisier, A. L. [1777] 1862. Réflexions sur le Phlogiston. In *Oeuvres de Lavoisier*. Paris: Imprimerie Impériale.

Lee, V. L. 1988. *Beyond Behaviorism*. Mahwah, NJ: Erlbaum.

1992. Transdermal Interpretation of the Subject Matter of Behavior Analysis. *American Psychologist* 47:1337–43.

Lennox, J. G. 1991. Darwinian Thought Experiments: A Function for Just-So Stories. In *Thought Experiments in Science and Philosophy*, ed. T. Horowitz and G. J. Massey. Savage, MD: Rowman and Littlefield, pp. 223–45.

Lenski, R. E., and Mittler, J. E. 1993. The Directed Mutation Controversy and Neo-Darwinism. *Science* 259:188–94.

Levin, S. G., Stephan, P. E., and Walker, M. B. 1995. Planck's Principle Revisited: A Note. *Social Studies of Science* 2:275–83.

Levitt, N. 1997. Deconstructing Science. *Science* 276:1953.

Lewin, R. 1985. Molecular Clocks Scrutinized. *Science* 228:571.

Lewontin, R. C. 1970. The Units of Selection. *Annual Review of Ecology and Systematics* 1:1–18.

Lightman, B. 1997. Introduction. In *Victorian Science in Context*, ed. B. Lightman. Chicago: University of Chicago Press.

Lipton, P. L., and Thompson, N. S. 1988. Comparative Psychology and the Recursive Structure of Filter Explanations. *International Journal of Comparative Psychology* 1:215–29.

Lisman, J. E., and Fallon, J. R. 1999. What Maintains Memories? *Science* 283:339–40.

Lloyd, E. A. 1988. *The Structure and Confirmation of Evolutionary Theory*. New York: Greenwood.

1992. Units of Selection. In *Keywords in Evolutionary Biology*, ed. E. Fox Keller and E. A. Lloyd. Cambridge, MA: Harvard University Press, pp. 334–40.

Loewenberg, B. 1932. The Reaction of American Scientists to Darwinism. *American Historical Review* 38:687–701.

Lumsden, C. J., and Wilson, E. O. 1981. *Genes, Mind and Culture: The Coevolutionary Process*. Cambridge, MA: Harvard University Press.

Lyell, C. 1881. *Life, Letters and Journals of Sir Charles Lyell, bart.*, ed. K. M. Lyell. London: Murray.

Macilwain, C. 1997a. "Science Wars" Blamed for Loss of Post. *Nature* 387:325.

1997b. Campuses Ring to a Stormy Clash over Truth and Reason. *Nature* 387:331–3.

MacIntyre, A. 1981. *After Virtue*. Notre Dame, IN: University of Notre Dame Press.

MacPhee, D. G., and Ambrose, M. 1996. Spontaneous Mutations in Bacteria: Chance or Necessity? *Genetica* 97:87–101.

MacRoberts, M. H., and MacRoberts, B. R. 1984. The Negational Reference: Or the Art of Dissembling. *Social Studies of Science* 14:91–3.

Mahoney, M. J. 1979. Psychology of the Scientist: An Evaluative Review. *Social Studies of Science* 9:349–75.

Margolis, L. 1996. Gaia Is a Tough Bitch. In *The Third Culture: Beyond the Scientific Revolution*, ed. J. Brockman. New York: Simon and Schuster, pp. 129–46.

Marshall, E. 1986. San Diego's Tough Stand on Research Fraud. *Science* 234:534–5.

Marx, J. 1995. How DNA Replication Originates. *Science* 270:1585–7.

Massey, G. J. 1991. Backdoor Analyticity. In *Thought Experiments in Science and Philosophy*, ed. T. Horowitz and G. J. Massey. Savage, MD: Rowman and Littlefield, pp. 285–96.

Masterson, J. 1994. Stomata Size in Fossil Plants: Evidence of Polyploidy in Majority of Angiosperms. *Science* 264:421–4.

Matthew, P. 1831. *On Naval Timber and Arboriculture*. London: Longman, Rees, Orme, Brown and Green.

Maynard Smith, J. 1971. The Origin and Maintenance of Sex. In *Group Selection*, ed. G. C. Williams. New York: Aldine-Atherton, pp. 163–75.

Mayr, E. 1963. *Animal Species and Evolution*. Cambridge, MA: Harvard University Press.

1983. Comments on David Hull's Paper on Exemplars and Type Specimens. In *PSA 1982*, vol. 2, ed. P. D. Asquith and T. Nickles. East Lansing, MI: Philosophy of Science Association, pp. 504–11.

1987. The Ontological Status of Species: Scientific Progress and Philosophical Terminology. *Biology & Philosophy* 2:145–66.

1997. The Objects of Selection. *Proceedings of the National Academy of Sciences* 94:2091–4.

McCann, H. G. 1978. *Chemistry Transformed*. Norwood, NJ: Ablex.

Medawar, P. B. 1972. *The Hope of Progress*. London: Methuen.

1977. Fear and DNA. *New York Review of Books*, October 27, pp. 15–20.

Meltzer, L. 1956. Scientific Productivity in Organizational Settings. *Journal of Social Issues* 12:32–40.

Menard, H. W. 1971. *Science: Growth and Change*. Cambridge, MA: Harvard University Press.

Mendel, G. 1865. Versuche über Pflanzen-hybriden. *Verhandlungen des Naturforschenden Veriens Brün* 4:3–47 (appeared in 1866).

Merton, R. K. 1968. The Matthew Effect. *Science* 159:56–63.

Merton, R. K., and Zuckermann, H. 1972. Age, Aging, and Age Structure in Science. In *A Sociology of Age Stratification*, ed. M. W. Riley, M. Johnson, and A. Foner. New York: Russell Sage Foundation. Reprinted (1973) in *The Sociology of Knowledge*, ed. Norman W. Storer. Chicago: University of Chicago Press, pp. 497–559.

Messeri, A. M. 1988. Age Differences in the Reception of New Scientific Theories: The Case of Plate Tectonics Theory. *Social Studies of Science* 18:91–112.

Michod, R. E. 1982. The Theory of Kin Selection. *Annual Review of Ecology and Systematics* 13:23–55.

Mill, J. S. 1843. *A System of Logic*. London: Longmans, Green.

Milliken, R. G. 1984. *Language, Thought, and Other Biological Categories: New Foundations for Realism*. Cambridge, MA: MIT Press.

Mirowski, P. 1994. A Visible Hand in the Marketplace of Ideas: Precision-Measurement as Arbitrage. *Science in Context* 7:563–89.

Mishler, B. D., and Donoghue, M. J. 1982. Species Concepts: A Case for Pluralism. *Systematic Zoology* 31:491–503.

Montgomery, W. M. 1974. German Editions of Important Works on Evolution. In *The Comparative Reception of Darwinism*, ed. T. Glick. Austin: University of Texas Press, pp. 81–116.

Muller, H. J. 1964. The Relation of Recombination to Mutational Advance. *Mutation Research* 1:2–9.

Nargeot, R., Baxter, D. A., and Bryne, J. H. 1997. Contingent-Dependent Enhancement of Rhythmic Motor Patterns: An In Vitro Analog of Operant Conditioning. *Journal of Neuroscience* 17:8093–105.

    1999. In Vitro Analog of Operant Conditioning in Aplysia. II. Modifications of the Functional Dynamics of an Identified Neuron Contribute to Motor Pattern Selection. *Journal of Neuroscience* 19:2261–72.

Neander, K. 1991. Functions as Selected Effects: The Conceptual Analyst's Defense. *Philosophy of Science* 58:168–84.

Nelson, G., and Platnick, N. 1981. *Systematics and Biogeography: Cladistics and Vicariance*. New York: Columbia University Press.

Nersessian, N. J. 1993. In the Theoretician's Laboratory: Thought Experimenting as Mental Modeling. In *PSA 1992*, vol. 2, ed. D. Hull, M. Forbes, and K. Okruhlik. East Lansing, MI: Philosophy of Science Association, pp. 291–301.

Newton-Smith, W. H. 1981. *The Rationality of Science*. London: Routledge and Kegan Paul.

Nitecki, M. N., Lemke, J. L., Pullman, H. W., and Johnson, M. E. 1978. Acceptance of Plate Tectonic Theory by Geologists. *Geology* 6:661–4.

Norman, C. 1984. Reduce Fraud in Seven Easy Steps. *Science* 224:581.

    1987. Prosecution Urged in Fraud Case. *Science* 236:1057.

Norton, J. D. 1993. A Paradox in Newtonian Gravitation Theory. In *PSA 1992*, vol. 2, ed. D. Hull, M. Forbes, and K. Okruhlik. East Lansing, MI: Philosophy of Science Association, pp. 414–22.

Nozick, R. 1974. *Anarchy, State, and Utopia*. New York: Basic Books.

Numbers, R. L. 1998. *Darwinism Comes to America: A Reevaluation of Scientific Responses*. Cambridge, MA: Harvard University Press.

Nunan, R. 1988. The Theory of an Expanding Earth and the Acceptability of Guiding Assumptions. In *Scrutinizing Science: Empirical Studies of Scientific Change*, ed. A. Donovan, L. Laudan, and R. Laudan. Dordrecht, Holland: Kluwer, pp. 289–314.

Oakley, K. P. 1964. The Problems of Man's Antiquity. *Bulletin of the British Museum (Natural History)* 9(5):1–155.

Odling-Smee, F. J., 1996. Niche Construction, Genetic Evolution, and Cultural Change. *Behavioural Processes* 35:195–205.

Odling-Smee, F. J., and Plotkin, H. C. 1984. Evolution: Its Levels and Its Units. *Behavioral and Brain Sciences* 7:318–20.

Oyama, S. 1985. *The Ontogeny of Information: Developmental Systems and Evolution*. Cambridge University Press.

    1989. Ontogeny and the Central Dogma. In *Systems and Development*, ed. M. R. Gunnar and E. Thales. Mahwah, NJ: Erlbaum.

References

Page, S., and Neuringer, A. 1985. Variability Is an Operant. *Journal of Experimental Psychology: Animal Behavior Processes* 11:429–52.

Parascandola, M. 1995. Philosophy in the Laboratory: The Debate over Evidence for E. J. Steele's Lamarckian Hypothesis. *Studies in History and Philosophy of Science* 26:469–92.

Paul, H. W. 1974. Religion and Darwinism: Varieties of Catholic Reaction. In *The Comparative Reception of Darwinism*, ed. T. Glick. Austin: University of Texas Press, pp. 403–36.

Pear, J. J., and Legris, J. A. 1987. Shaping by Automated Tracking of an Arbitrary Operant Response. *Journal of the Experimental Analysis of Behavior* 47:241–7.

Peck, J. R., and Eyre-Walker, A. 1998. The Muddle about Mutations. *Nature* 387:135–6.

Pendlebury, D. 1991. Science, Citation, and Funding. *Science* 251:1410–1.

Pennisi, E. 1998. How the Genome Readies Itself for Evolution. *Science* 281:1131–4.

Perrin, C. E. 1988. The Chemical Revolution: Shifts in Guiding Assumptions. In *Scrutinizing Science*, ed. A. Donovan, R. Laudan, and L. Laudan. Dordrecht, Holland: Kluwer, pp. 105–24.

Pettit, P. 1996. Functional Explanation and Virtual Selection. *British Journal for the Philosophy of Science* 47:291–302.

Piattelli-Palmarini, M. 1986. The Rise of Selection Theories: A Case Study and Some Lessons from Immunology. In *Language, Learning and Concept Acquisition: Foundational Issues*, ed. W. Demopoulos and A. Marras. Norwood, NJ: Ablex, pp. 117–30.

Planck, M. 1936. *The Philosophy of Physics*. New York: Norton.

1949. *Scientific Autobiography and Other Papers*, trans. F. Gaynor. New York: Philosophical Library.

Plotkin, H. C. 1987. Evolutionary Epistemology as Science. *Biology & Philosophy* 2:87–105.

1991. The Testing of Evolutionary Epistemology. *Biology & Philosophy* 6:481–97.

1994. *The Nature of Knowledge*. New York: Penguin.

Plotkin, H. C., and Odling-Smee, F. J. 1981. A Multiple-Level Model of Evolution and Its Implications for Sociobiology. *Behavioral and Brain Sciences* 4:225–35.

Pocklington, R., and Best, M. L. 1997. Cultural Evolution and Units of Selection in Replicating Text. *Journal of Theoretical Biology* 188:79–87.

Popper, K. R. 1962. *Conjectures and Refutations*. New York: Basic Books.

1972. *Objective Knowledge*. Oxford: Clarendon.

Price, D. de S. 1963. *Little Science, Big Science*. New York: Columbia University Press.

Putnam, H. 1973. Meaning and Reference. *Journal of Philosophy* 7:699–711.

1981. *Reason, Truth and History*. Cambridge University Press.

Quartz, S. R., and Sejnowski, T. J. 1997. The Neural Basis of Cognitive Development: A Constructivist Manifesto. *Behavioral and Brain Sciences* 20:537–96.

References

Quillian, M. R. 1994. A Content-Independent Explanation of Science's Effectiveness. *Philosophy of Science* 61:429–48.

Rappa, M., and Debackere, K. 1993. Youth and Scientific Innovation: The Role of Young Scientists in the Development of a New Field. *Minerva* 31:1–20.

Rawling, A. 1994. The AIDS Virus Dispute: Awarding Priority for the Discovery of the Human Immunodeficiency Virus (HIV). *Science, Technology & Human Values* 19:342–60.

Ray, B. A., and Sidman, M. 1970. Reinforcement Schedules and Stimulus Control. In *Theory of Reinforcement Schedules*, ed. W. N. Schoenfeld. New York: Appleton-Century-Crofts, pp. 187–214.

Rescher, N. 1991. Thought Experiments in Presocratic Philosophy. In *Thought Experiments in Science and Philosophy*, ed. T. Horowitz and G. J. Massey. Savage, MD: Rowman and Littlefield, pp. 31–41.

Reskin, B. F. 1977. Scientific Productivity and the Reward Structure of Science. *American Sociological Review* 42:491–504.

Richards, E., and Ashmore, M. 1996. More Sauce Please! The Politics of SSK: Neutrality, Commitment and Beyond. *Social Studies of Science* 26:219–28.

Richards, R. J. 1993. History as the Necessary Foundation for Philosophy of Science. In *PSA 1992*, vol. 2, ed. D. Hull, M. Forbes, and K. Okruhlik. East Lansing, MI: Philosophy of Science Association, pp. 482–9.

Richerson, P. J., and Boyd, R. 1987. Simple Models and Complex Phenomena: The Case of Cultural Evolution. In *The Latest on the Best: Essays on Evolution and Optimality*, ed. J. Dupré. Cambridge, MA: MIT Press.

Ridley, M. 1983. *The Explanation of Organic Diversity: The Comparative Method and Adaptations for Mating*. Cambridge, MA: Harvard University Press.

Ritland, D. B., and Brower, L. P. 1991. The Viceroy Butterfly Is Not a Batesian Mimic. *Science* 350:497–8.

Robinson, J. D. 1997. Deconstructing Science. *Science* 276:1953.

Rosenberg, A. 1985. *The Structure of Biological Science*. Cambridge University Press.

1989. From Reductionism to Instrumentalism? In *What the Philosophy of Biology Is: Essays Dedicated to David Hull*, ed. M. Ruse. Dordrecht, Holland: Kluwer, pp. 245–62.

Rosenberg, S. M., Harris, R. S., and Torkelson, J. 1995. Molecular Handles on Adaptive Mutation. *Molecular Microbiology* 18:185–9.

Rosenkrantz, R. D. 1977. *Inference, Method and Decision*. Dordrecht, Holland: Reidel.

Rudwick, M. J. S. 1985. *The Great Devonian Controversy: The Making of Scientific Knowledge among Gentlemanly Specialists*. Chicago: University of Chicago Press.

Ruse, M. 1975. Charles Darwin and Artificial Selection. *Journal of the History of Ideas* 36:339–50.

1979. *The Darwinian Revolution: Science Red in Tooth and Claw*. Chicago: University of Chicago Press.

1989a. David Hull through Two Decades. In *What the Philosophy of Biology Is: Essays Dedicated to David Hull*, ed. M. Ruse. Dordrecht, Holland: Kluwer, pp. 1–15.

References

1989b. The View from Somewhere: A Critical Defense of Evolutionary Epistemology. In *Issues in Evolutionary Epistemology*, ed. K. Halweg and C. A. Hooker. Albany: State University of New York Press, pp. 185–228.

ed. 1989c. *What the Philosophy of Biology Is: Essays Dedicated to David L. Hull.* Dordrecht, Holland: Kluwer.

1996. *Monad to Man: The Concept of Progress in Evolutionary Biology.* Cambridge, MA: Harvard University Press.

Salmon, W. C. 1998. *Causality and Explanation.* Oxford: Oxford University Press.

Samuelson, P. 1966. The General Theory. In *The Collected Papers of Paul Samuelson*, ed. J. Stiglitz. Cambridge, MA: MIT Press.

Sandler, N. 1997. Deconstructing Science. *Science* 276:1954–5.

Sarkar, H. 1982. A Theory of Group Rationality. *Studies in the History and Philosophy of Science* 1:55–72.

1996. Biological Information: A Skeptical Look at Some Central Dogmas of Molecular Biology. In *The Philosophy and History of Molecular Biology: New Perspectives*, ed. S. Sarkar. Dordrecht, Holland: Kluwer, pp. 187–231.

Scheffler, I. 1967. *Science and Subjectivity.* New York: Bobbs-Merrill.

Schilcher, F., and Tennant, N. 1984. *Philosophy, Evolution and Human Nature.* London: Routledge and Kegan Paul.

Schwartz, B. 1974. On Going Back to Nature: A Review of Seligman and Hager's "Biological Boundaries of Learning." *Journal of the Experimental Analysis of Behavior* 21:183–98.

Searle, J. R. 1984. Intentionality and Its Place in Nature. *Dialectica* 38:86–99.

Self, D. W., and Stein, L. 1992. Receptor Subtypes in Opioid and Stimulant Reward. *Pharmacology & Toxicology* 701:87–94.

Semon, R. 1904. *Die Mneme als erhaltendes Prinzip in Weschsel des organischen Geschehens.* Leipzig: Engelmann.

Sen, A. K. 1983. The Profit Motive. *Lloyds Bank Review* 147 (January):1–20.

Sidman, M. 1994. *Equivalence Relations and Behavior: A Research Story.* Boston: Authors Cooperative.

Signor III, P. W. 1985. Real and Apparent Trends in Species Richness Through Time. In *Phanerozoic Diversity: Profiles in Macroevolution*, ed. J. Valentine. Princeton: Princeton University Press, pp. 129–50.

Silverstein, A. M., and Rose, N. R. 1997. On the Mystique of the Immunological Self. *Immunology Review* 159:197–206.

Skinner, B. F. 1953. *Science and Human Behavior.* New York: Free Press.

1974. *About Behaviorism.* New York: Knopf.

1981. Selection by Consequences. *Science* 213:501–4.

1984. The Evolution of Behavior. *Journal of Experimental Analysis of Behavior* 41:217–21.

Smith, A. [1776] 1993. *An Inquiry into the Nature and Causes of the Wealth of Nations*, abridged with commentary and notes by L. Dickey. Indianapolis: Hackett.

Sneath, P. H. A., and Sokal, R. R. 1973. *Numerical Taxonomy: The Principles and Practice of Numerical Classification.* San Francisco: Freeman.

References

Sober, E. 1981. Holism, Individualism, and the Units of Selection. In *PSA 1980*, vol. 2, ed. R. Asquith and R. Giere. East Lansing, MI: Philosophy of Science Association, pp. 93–121.

1984. *The Nature of Selection*. Cambridge, MA: MIT Press.

1989. Systematics and Circularity: Theory. In *What the Philosophy of Biology Is: Essays Dedicated to David Hull*, ed. M. Ruse. Dordrecht, Holland: Kluwer, pp. 263–73.

1992. Screening-off and the Units of Selection. *Philosophy of Science* 59:142–52.

Sober, E., and Lewontin, R. C. 1982. Artifact, Causes, and Genic Selection. *Philosophy of Science* 49:147–76.

Sober, E., and Wilson, D. S. 1998. *Unto Others: The Evolution and Psychology of Unselfish Behavior*. Cambridge, MA: Harvard University Press.

Sokal, A. D. 1996a. Transgressing Boundaries: Toward a Transnormative Hermeneutics of Quantum Gravity. *Social Text* 14:217–52.

1996b. A Physicist Experiments with Cultural Studies. *Lingua Franca*, May/June, pp. 62–4.

Solomon, M. 1995. Multivariate Models of Scientific Change. In *PSA 1994*, vol. 2, ed. D. L. Hull, M. Forbes, and R. M. Burian. East Lansing, MI: Philosophy of Science Association, pp. 287–97.

1996. Information and the Ethics of Information Control in Science. *Perspectives on Science* 4:195–206.

Sorenson, R. 1992. *Thought Experiments*. Oxford: Oxford University Press.

Staddon, J. E. R., and Simmelhag, V. L. 1971. The "Superstition" Experiment: A Reexamination of Its Implications for the Principles of Adaptive Behavior. *Psychological Review* 78:3–43.

Stegmuller, W. 1976. *The Structure and Dynamics of Theories*. New York: Springer-Verlag.

Stein, L., and Belluzzi, J. D. 1988. Operant Conditioning of Individual Neurons. In *Quantitative Analyses of Behavior*, vol. 7, ed. M. L. Commons, R. M. Church, J. R. Stellar, and A. R. Wagner. Hillsdale, NJ: Erlbaum, pp. 249–64.

Stein, L., Xue, B. G., and Belluzzi, J. D. 1994. In Vitro Reinforcement of Hippocampal Bursting: A Search for Skinner's Atoms. *Journal of the Experimental Analysis of Behavior* 61:155–68.

Stephan, P. E., and Levin, S. G. 1992. *Striking the Mother Lode in Science*. Oxford: Oxford University Press.

Stone, J. R. 1996. The Evolution of Ideas: A Phylogeny of Shell Models. *American Naturalist* 148:904–29.

Sulloway, F. J. 1996. *Born to Rebel: Birth Order, Family Dynamics and Creative Lives*. New York: Pantheon.

Taubes, G. 1997. Computer Design Meets Darwin. *Science* 277:1931–2.

Thomas, L. 1974. Survival by Self-Sacrifice. *Harper's Magazine* 251:96–104.

Thomason, S. G. 1991. Thought Experiments in Linguistics. In *Thought Experiments in Science and Philosophy*, ed. T. Horowitz and G. J. Massey. Savage, MD: Rowman and Littlefield, pp. 247–57.

Thompson, J. J. 1971. A Defense of Abortion. *Philosophy & Public Affairs* 1:47–66.

References

Thompson, M. S. 1994. Vehicles All the Way Down. *Behavioral and Brain Sciences* 17:638.

Todd, J. T., and Morris, E. K. 1992. Case Histories in the Great Power of Steady Misrepresentation. *American Psychologist* 47:1441–53.

Toulmin, S. 1972. *Human Understanding*. Princeton: Princeton University Press.

Trefil, J. 1997. Deconstructing Science. *Science* 276:1953–4.

Ullmann-Margalit, E. 1978. Invisible-Hand Explanations. *Synthese* 39:263–92.

van der Steen, W. 1996. Screening-off and Natural Selection. *Philosophy of Science* 632:115–21.

Vrba, E. S. 1984. What Is Species Selection? *Systematic Zoology* 33:318–28.

Vrba, E. S., and Eldredge, E. 1984. Individuals, Hierarchies, and Processes: Towards a More Complete Evolutionary Theory. *Paleobiology* 10:146–71.

Vrba, E. S., and Gould, S. J. 1986. The Hierarchical Expansion of Sorting and Selection: Sorting and Selection Cannot Be Equated. *Paleobiology* 12:217–28.

Wade, N. 1981. *The Nobel Duel*. Garden City, NY: Anchor.

Waters, C. K. 1986. Taking Analogical Inferences Seriously: Darwin's Argument from Artificial Selection. In *PSA 1986*, vol. 1, ed. A. Fine and P. Machamer. East Lansing, MI: Philosophy of Science Association, pp. 501–13.

1991. Tempered Realism about the Force of Selection. *Philosophy of Science* 58:553–73.

Webster, G., and Goodwin, B. 1996. *Form and Transformation*. Cambridge University Press.

Weintraub, E. R., and Mirowski, P. 1994. The Pure and the Applied: Bourbakism Comes to Mathematical Economics. *Science in Context* 7:245–72.

Whewell, W. 1837. *History of the Inductive Sciences*. London: J. W. Parker.

1849. *Of Induction, with special reference to Mr. J. Stuart Mill's System of Logic*. London: J. W. Parker.

1851. On the Transformation of Hypotheses in the History of Science. *Transactions of the Cambridge Philosophical Society* 9:139–47.

White, R. R. 1987. Accuracy and Truth. *Science* 325:1447.

Wiggins, D. 1967. *Identity and Spatio-Temporal Continuity*. Oxford: Blackwell.

Wilkes, K. 1988. *Real People: Personal Identity without Thought Experiments*. Oxford: Oxford University Press.

Williams, G. C. 1966. *Adaptation and Natural Selection*. Princeton: Princeton University Press.

ed. 1971. *Group Selection*. New York: Aldine-Atherton.

1975. *Sex and Evolution*. Princeton: Princeton University Press.

1985. A Defense of Reductionism. *Oxford Surveys in Evolutionary Biology* 2:1–27.

1992. *Natural Selection: Domains, Levels, and Challenges*. Oxford: Oxford University Press.

1996. Reply to Johnson. *Biology & Philosophy* 11:541.

Wilson, D. S., and Sober, E. 1994. Re-introducing Group Selection to the Human Behavioral Sciences. *Behavioral and Brain Sciences* 17:585–654.

Wilson, E. O. 1975. *Sociobiology: The New Synthesis*. Cambridge, MA: Harvard University Press.

References

Wimsatt, W. 1980. Reductionist Research Strategies and Their Bases in the Units of Selection Controversy. In *Scientific Discovery*, ed. T. Nickles. Dordrecht, Holland: Reidel, pp. 213–59.

——— 1981. Units of Selection and the Structure of the Multi-Level Genome. In *PSA 1980*, vol. 2, ed. P. D. Asquith and T. Nickles. East Lansing, MI: Philosophy of Science Association, pp. 122–83.

Wisdom, J. O. 1974. The Nature of "Normal" Science. In *The Philosophy of Karl Popper*, vol. 2, ed. P. A. Schilpp. La Salle, IL: Open Court, pp. 820–42.

Wolpert, L. 1992. *The Unnatural Nature of Science*. Cambridge, MA: Harvard University Press.

Woolgar, S. 1988. *Science: The Very Idea*. London: Tavistock.

Ylikoski, P. 1995. The Invisible Hand and Science. *Science Studies* 8:32–43.

Yoels, W. C. 1973. On "Publishing or Perishing": Fact or Fable? *American Sociologist* 8:128–34.

Young, R. M. 1971. Darwin's Metaphor: Does Nature Select? *Monist* 55:442–503.

Ziegler, C. A. 1997. Deconstructing Science. *Science* 276:1955.

Ziman, J. 1987. *Knowing Everything About Nothing*. Cambridge University Press.

——— 1994. *Prometheus Bound: Science in a Dynamic Steady State*. Cambridge University Press.

Zuckerman, H. 1977. Deviant Behavior and Social Control in Science. In *Deviance and Social Change*, ed. E. Sagarin. Beverly Hills: Sage, pp. 87–138.

Index

Index